The Critical Chain Implementation Handbook

By David Updegrove

The Critical Chain Implementation Handbook

ISBN-13: 978-1499576733
ISBN-10: 1499576730

"If I have seen a little further, it is by standing on the shoulders of Giants."
Sir Isaac Newton

For Eli...

This book is dedicated to the memory of Dr. Eliyahu M. Goldratt,

Teacher, mentor, friend, GIANT

Having the opportunity to work closely with Eli is one of the greatest

honors of my life.

Acknowledgements

Have you ever noticed how most authors include their family in the acknowledgements, but put them in dead last? Sometimes they qualify it with a "last but not least," or "my source of strength," or some similar platitude, but the plain truth is without my wife I would not have the knowledge to write this book. She was the one who encouraged me in early 2000 to take a huge risk, leaving a 20-year career as an engineer at Boeing, and becoming a TOC consultant in the wider world. This allowed me to gather experience from dozens of companies in every corner of the globe, and eventually to work with Eli Goldratt himself. For this I am forever grateful to my beautiful wife, Tess.

Hilbert Robinson has been a friend and confidant for many years, and I consider him one of the elite implementers of Critical Chain on the planet. I will always be thankful for the (continuing to this day) countless hours of discussion on every aspect of Critical Chain, many of which took place in Hawaii while we worked together at Pearl Harbor. Hilbert later became a founder, along with two others and myself, of our own TOC consulting company. Hilbert was also one of my main readers and contributed a tremendous amount to my manuscript.

I have similar high regard for Dani Omri, with whom I worked for nearly two years in Israel and India. I learned many invaluable lessons on Critical Chain and life itself from Dani's knowledge, wisdom, and humanity. Much of the content of this book comes from my experience with Dani. He was also one of my main readers, and his contribution is stamped all over the book, for example with the concept of thousands of tiny random "freezes" in the traditional multi-project environment. He also contributed important thoughts to Chapter 30.

Thanks to Rodger Morrison, Brad Cartier (the two others from my consulting company), Doug Krsek, Sundara Raghavan, Pranav Bhosekar and many others, with whom I have spent time in the implementation trenches. They always challenged my assumptions and prompted me to think.

Many people have helped me to grow in the Thinking Processes, Marketing and Sales, and the most effective bedside manner for a consultant. Notable among them are Michael Demere, who joined me in my venture to distribute Critical Chain software. Michael always kept me on my toes, never allowed me to be lazy, and always challenged me to *think*.

Two others whom I consider close friends and advisers are Yaniv Dinur and Amir Davidi, whom I worked with while with Goldratt Consulting in Israel. Both of them are amazing in their knowledge and application of TOC, as well as the ability to communicate it to others.

I have deep respect and gratitude for my Japanese partners at Being Company, Ltd. in Japan, who have created a superior CCPM software product called BeingManagement3. Being's president, Masahiro Suehiro, and Takashi Nakamura, who was formerly in charge of international business offered me great support. Ichigoro Kuroki, head of Japan's TOC Think Tank, is the primary translator of the Japanese version of this book, and for that he has my warmest thanks. Thanks also to the brilliant former chief developer of BeingManagement software, Koichi Ujigawa, who is now in charge of all TOC activity at Being. But especially important and inspirational to me is Yoshishige Tsuda, founder and Chairman of the Board of Being. If I can achieve becoming the same kind of leader he is, I will have done well indeed.

I am grateful to Rami Goldratt for giving me the opportunity to work with himself and his father. I will always remember Rami for giving me an exciting one-hour tour of Jerusalem (including a meal)! Now that was an effective, short duration project!

Thanks to Goldratt Research Labs for permission to use excerpts from the Projects Company Strategy & Tactics tree as depicted in their software, the "Harmony TOC S&T Expert System."

Thanks to V.V. Risbud and B. Subramanyam of Larsen & Toubro Ltd. in India for the use of project freeze and full kitting photographs. I wish you nothing but success.

Finally, I would like to acknowledge Lisa Scheinkopf, from whom I gathered the details of the genesis of Critical Chain in her great 2011 TOCICO presentation.

IMPORTANT NOTE:

Several versions of the generic Projects S&T Template exist today. For this book I have used the August, 2010 version. Since this version is the last version published while Dr. Goldratt was still alive, I consider this the authoritative version until a TOC-community consensus-based body is established to consider changes/additions to the template.

JOIN THE DISCUSSION:

Find additional materials, ask questions, share your ideas, and discuss The Critical Chain Implementation Handbook on Facebook: http://www.facebook.com/ImplementingCCPM. Or when in Facebook, do a search on The Critical Chain Implementation Handbook.

TABLE OF CONTENTS

Section One: Setting the Table

Flow is the number one consideration

Introduction: Building the High-Performance Projects Organization

Flow is the number one consideration.

In 1986, representatives from the Norwegian energy company Statoil approached Dr. Eli Goldratt with a proposal. As builders of North Sea oil rigs, the people at Statoil had by necessity become project management experts. In fact, due to the nature of oil rig construction, Statoil's expertise in project management could be argued to have been the best in the world. Yet in spite of this expertise, they still had great difficulty in completing their projects on time. They told Goldratt that to really determine the duration of a project, they would multiply the planned duration of the project by 4 and pray. The problem, they told him, was that their prayers weren't answered. They asked him for help solving their dilemma.

Goldratt protested that his experience was with production, not project management. But Statoil's representatives replied that although this was understood, they had read *The Goal*[1], and were convinced that if anyone could figure out a solution for project management, it would be Eli Goldratt. They gave him two small booklets containing, they said, everything they knew about project management. Goldratt learned that lead time was the key for projects, and that lead time was determined by the Critical Path, a concept first pioneered in the late 1950s. Additionally, he learned that in one of their projects – originally comprised of 40,000 tasks – the Critical Path actually only totaled about 40 tasks, or 1/10th of 1% of the project. The idea that the overall duration of a project is determined by a very small number of tasks captured his attention, and he agreed to work on the problem.

Whether it was production or projects, Goldratt realized everything revolved around constraints. In production, the constraint was capacity-based. In projects, where the objective is usually to minimize time, the constraint is the sequence of tasks that determine how long the project

[1] Eliyahu M. Goldratt and Jeff Cox, North River Press, 1984

will take: the Critical Path. As a result of his visit to Statoil, and this second application of his research, the term "Theory of Constraints" was coined.

Two weeks later Goldratt boarded an airplane for Oslo. Unfortunately, in spite of the fact he had come to these understandings, he had yet to construct a solution. What would he tell Statoil? He realized there must be a fundamental assumption he had missed – the thing that was devouring the available time. As the 5-hour flight from Tel Aviv began, he started going over the situation in his head once again. His thought process went something like this:

1) *Critical path provides companies with a fantastic ability to focus. Nevertheless they are chronically late, even when they multiply by 4 and pray!*
2) *These are not stupid people, and Statoil in particular is the best in the world at projects! Their estimates must be, on average, good enough... How could it be that multiplying by such a large factor still does not provide enough time?*
3) *There must be a fundamental assumption that is missed, which is devouring the time...*
4) *The duration of a project is the sum of the duration of the tasks along its critical path...*
5) *But the actual duration of any task varies – often quite significantly – from the estimate...*
6) *When a task on the critical path takes longer than its estimate, the whole project takes longer...*
7) *But when a task on the critical path takes less time than its estimate, the project should be shorter. But this is apparently not happening! Why?*
8) *Show me how you'll measure me, and I'll show you how I'll behave...*
9) *If estimated task durations tend to become committed task durations, peoples' estimates will be inflated to contain enough "safety"...*
10) *If I am a task manager, since I have lots of safety and there are other urgent tasks, I can afford to start later...*
11) *If things magically go well and I finish early and report I'm done, I'll lose my safety for next time, when things might not go so well. So I'd better not report an early finish...*

12) *If I as a task manager have nothing else to do, and if being idle is frowned upon, I need to look busy—so I'll pace myself to make sure I don't finish early...*
13) *This explains why only delays accumulate...*
14) *And in projects there are many dependencies and variables... so these behaviors happen over and over again...*

As the plane neared Oslo, the answer came to him—the first major premise of Critical Chain:

15) *Remove safety from tasks and aggregate the safety into a project buffer and feeding buffers! Use the safety only when you need it!*

Over the next eleven years, Goldratt expanded on, tested and refined his solution. By 1997 it was mature enough to be given its own name and be put into the public domain. He published the book, *Critical Chain*[2], which was a business novel like *The Goal* and *It's Not Luck*.[3] It captured the attention of many, including myself, then an industrial engineer at Boeing. Soon thereafter I recommended Critical Chain be used to manufacture wing and tail assemblies for the F-22 fighter aircraft. My suggestion was accepted, and the people who implemented it got impressive results—so much so that the US Air Force team overseeing the contract asked Boeing what in the world they were doing to get such great delivery performance, and upon hearing the answer paid for the prime contractor team at Lockheed Martin to be trained in Critical Chain!

Both the book "Critical Chain" and its early implementations were mostly focused on the planning and execution of single projects. It soon became clear to most of us attempting Critical Chain that we were actually doing many (multiple) projects in parallel—the reality was we lived in a multi-project world. In fact, we realized that most organizations were actually *multi-project organizations*, executing many projects concurrently, with a shared workforce that often moved people from one project to another and then back again.

This was one of the most important insights to come out of the early attempts to implement Critical Chain. Failure to appreciate this

---

[2] Eliyahu M. Goldratt, North River Press, 1997

[3] Eliyahu M. Goldratt, North River Press, 1994

important distinction is a major reason that even today, many organizations are unable to grasp the full potential Critical Chain represents. They are unable to re-conceptualize the organization as a stable system through which projects constantly flow, optimally using the resources they need on demand. Instead, they are mentally anchored to the concept of independently funded, staffed and managed projects that take too long, while competing with each other for scarce resources. Matrix Management and Portfolio Management are well-meaning attempts to resolve the problem of sharing resources across multiple projects. However in practical use, they do not effectively provide fluid and effective transition of resources between projects.

During the early years of Critical Chain, it was clear that we had found a better way to deliver isolated projects on time. There still remained the question of how to deliver *all of them* on time—on budget—and with their full, originally intended scope. How could organizations ensure they weren't overloading their resources? What were the significant differences between single projects and multiple projects? As usual, Dr. Goldratt was already far ahead of the rest of us and had been testing the multi-project solution for a number of years.

Today the multi-project solution has been used to improve performance in new product development, software development, engineering-to-order, major assembly production, Maintenance, Repair & Overhaul (MRO), and many other environments in hundreds of implementations around the world. During this period it has matured and changed. The Resource Buffer concept has been abandoned, and the Projects Strategy & Tactics (S&T) tree has been created. Consultants have added Critical Chain to their menu of client offerings, and the Theory of Constraints International Certification Organization (TOCICO) conducts exams to certify Critical Chain Project Management (CCPM) experts. Software companies have created applications to aid implementation, and books have been written to further the knowledge of what Critical Chain is and how it works. Many implementations have been successful in reducing the duration of projects, cutting costs, and bringing back sanity and order to the workplace. Some have failed. Some have flourished for a while, but then have slowly faded until they are finally abandoned or unceremoniously snuffed out by the immune system of stagnating host organizations.

What accounts for this? Why does such a powerful concept produce some implementations that work spectacularly and endure for years, while others fail or disintegrate over time? What are the key elements of success? What are the potential pitfalls and hazards? Is there a straightforward way to help insure that an organization will achieve the first scenario, rather than experience the less desirable outcomes latter two? Is there a step-by-step method of implementing that can act as a roadmap to our success? I believe there is, and that is the purpose of this book.

It is my claim that Dr. Goldratt has captured all the necessary conditions for multi-project organization success in his Projects Strategy & Tactics (S&T) tree. Therefore the S&T tree provides the basic structure for this book. When followed precisely, the implementing organization has an excellent chance of successfully building, capitalizing upon, and sustaining a High Performance Projects Organization. When anything is compromised or left out, the efficacy of the implementation quickly declines. Experience has shown that despite the availability of the Projects Strategy & Tactics (S&T) tree since late 2006, organizations continue to struggle with their attempts to implement Critical Chain. I believe there is a need to further explain how to use the Projects Strategy & Tactics (S&T) tree successfully. This book attempts to provide the reader with the benefit of the author's experience in using the S&T to successfully implement critical chain. It follows the S&T tree step by step, being careful to add clarification to important terms and concepts as well as providing additional commentary on important practical considerations.

So how did we have wildly successful implementations prior to the publishing of the S&T tree? In part, good consultants intuitively and logically understood and implemented the majority of the necessary conditions. We also had a lot of good luck. In retrospect, when there was failure, it is likely that a post-mortem analysis of the implementation would surface omissions, often many omissions, from the strategies and tactics delineated in the S&T tree. In fact, and I cannot stress this enough, the steps prior to network construction are absolutely crucial to success. When any of these are omitted or watered-down the success of the implementation is significantly imperiled.

A word of caution. This is not the book to read if you are looking for an explanation of Critical Chain concepts. There are several other books, including the TOC Handbook, which does an excellent job of that. Rather, you will find that this book offers much more practical and *practicable* step-by-step instruction in on "how-to" implement Critical Chain. I think you will find the tone of the book to be informal and conversational.

This book is for the Critical Chain Implementer who has studied the S&T tree. Moreover, it is a book for the implementation team, and even for the entire implementing organization. By providing good understanding of not only the logic of the solution, but also the logic of its implementation, implementers and the organizations they serve should experience more consistently successful implementations. In turn, successful implementations should result in the establishment of more and more High Performance Projects Organizations.

Finally, I don't think it will take you very long to notice that in the spirit of the S&T tree, a major theme throughout this book is:

Flow is the number one consideration.

I will repeat it until I am blue in the... er... typing fingers.

Flow is the number one consideration. It's not the only consideration, but it has to be the number one consideration at all times. Greatly accelerating the rate of flow of projects is the most important aspect of creating the high performance organization. Everything in the Projects S&T tree, and therefore by extension this book, is based on the premise that flow is the number one consideration. Every step is designed to increase flow. Every step is necessary if you are to increase flow to its maximum extent. Therefore if you depart from the instructions herein to subordinate to something other than flow (more on this below) or to do CCPM "the (insert your company's name here) way," your potential for improvement will be diminished if not eliminated entirely.

It should be understood that the idea of flow being the number one consideration is not unique to TOC or Critical Chain. For the truly great leaders in the history of manufacturing, flow was just as importantly the number one consideration. For Taiichi Ohno, the father of the Toyota Production System, it was all about flow. In addition, Ohno credited none other than Henry Ford with his ideas about flow. From Goldratt's excellent article *Standing on the Shoulders of Giants*[4] comes this excerpt:

"Like Ford, Ohno's primary objective was improving flow – decreasing lead time – as indicated in his response to the question about what Toyota is doing:

All we are doing is looking at the time line from the moment the customer gives us an order to the point when we collect the cash. And we are reducing that time line…" (Footnote: "Ohno, Taiichi, *Toyota Production System*, Productivity, Inc. 1988, page ix (in Publisher's forward).
It is also worth noting that in this and his other books Ohno gives full credit to Ford for the underlying concepts.)"

Decades earlier, in 1926, Henry Ford had made the following statement:

"One of the most noteworthy accomplishments in keeping the price of (our) products low is the gradual shortening of the production cycle. The longer an article is in the process of manufacture and the more it is moved about, the greater is its ultimate cost."

With names like Ford, Ohno, and Goldratt all proclaiming, in clear intent if not in these exact words that flow is the number one consideration, perhaps it would be wise for us to take note and behave accordingly. If you are embarking on implementing Critical Chain and you are struggling with the idea that you have to make flow the number one consideration in order to have a successful implementation, then I suggest that you step back and rethink your interest in Critical Chain.

A few years ago, while teaching Critical Chain at a large company, sort of as a joke but also to make a point, I was stressing that flow was the number one consideration, and that every morning, to remind them of this, they should not just say "good morning" to each other, but rather, "Good morning! Flow is the number one consideration!" To my surprise,

---

[4] Eliyahu M. Goldratt, Goldratt Consulting White Paper, 2008

when I came in the next day and said good morning to the group, I was greeted with a chorus of, "Good morning! Flow is the number one consideration!" Even weeks later, when I would pass someone in the hall or between buildings, I would get the same greeting. I repeated this with other clients and got the same effect. Of course the intent here is not to create a meaningless rote addition to our vocabulary, but to guide people toward the true internalization of the concept of flow being the number one consideration.

Yet this is not only a phrase to remind us, it is also a practical way to provide clear guidance for day-to-day decision making – similar to the role the North Star played for night travelers long ago. Here is how it works. Whenever a decision that will affect how we execute work is to be made—"Shall we do A or B?," we can more easily come to the best decision if we are all in agreement *a priori* that flow is to be our number one consideration. For example, "A" saves an inspector from walking to our building, but "B" shaves two days off the project." The obvious answer in this case is to choose "B". The inspector has to walk!

I stated above that although flow needs to be our number one consideration, it is not our only consideration. There may be other important considerations, such as safety, cost, or governmental compliance. But all of these should be examined with flow being the primary, or number one consideration. Even so, especially when it comes to cost, flow should win the vast majority of the time. This is because if minimizing cost is made the number one consideration you can unintentionally make decisions that impede flow and actually drive costs upward.

"We have to save cost, so we have a no new-hiring policy." Terrific. But if you make money by completing projects and you have a constraint in your organization that could be eliminated by hiring someone—increasing flow, delivering more and faster projects and thereby increasing the rate of money coming into the company, and you decide not to hire due to cost, then you also have a "no improvement policy," or even a "close our doors soon" policy.

Let me say it as clearly and as strongly as I can: **Critical Chain is the most effective cost savings and profit-generating device for the Projects organization.** Even from the cost point of view flow should be your number one consideration, better cost savings cannot be achieved than through realizing much faster flow. Remember Henry Ford's proclamation.

We need to be ever-vigilant concerning the things to which we subordinate. The no new-hiring policy? A batching policy? An ERP system? Cost accounting policies? Local efficiencies? By definition all companies subordinate to something. If flow is your number one consideration, it's easy to ask yourself, "What are we subordinating to here?" If it's anything other than moving projects faster through your organization, then you have a problem if you want to become a high performance—and high profit—organization. I am aware that there are some companies who make so much money today despite their ineffective operating strategy that the idea of improving the way they manage operational logistics is a low priority for them. Although I am certain they would make even more money from adopting Critical Chain, something tells me they might not be buying many copies of this book.

With that, let's get started. My hope is that this book helps your company to prosper, even to flourish. Having seen it many times now, there is no doubt in my mind that it can be done, and that as unique as your organization may be, it can be done with you. Best wishes in your Critical Chain implementation endeavors.

Oh, and one more thing: Flow is the number one consideration!

How To Use This Book

The *Critical Chain Implementation Handbook* is intended to be a genuine handbook (or field manual), to be used to train in preparation for implementation, as a roadmap during the implementation itself, and as an authority to evaluate the quality of the implementation post-execution. I actually expect it to be carried throughout the workplace and to different types of meetings by members of the implementation team as a frequent reference to keep the implementation on-track. I expect it to be an authoritative back-up for keeping top management informed and focused. I expect lots of margin notes and dog-eared pages. I also expect frequent deep discussions over the "three questions," (as explained in Chapter 4) key elements of implementation, vital for genuine progress through the Strategy and Tactics (S&T) tree, (which will be discussed at length in Chapter 2).

Moreover, this book is to be used post-implementation as a refresher to help sustain the implementation over time. As new employees come into the organization and experienced veteran employees retire and move on, the Handbook can help guard against the deterioration of knowledge and conviction and therefore the implementation.

In a moment I will convey a few warnings I need to share about using this book to implement Critical Chain. But before that there is one understanding that I think has a great amount of significance for properly applying the steps herein. I believe it is vital that we understand the distinction between planning and execution. For example, in *planning* we identify the Critical Chain of the various projects to find the fastest ways to complete them. In *execution*, the individual Critical Chains virtually vanish, and we assign resources and prioritize tasks based on the degree of buffer penetration across projects. In *planning* we plan optimal resources for all tasks in order to execute tasks as fast as we think is possible. In *execution* we re-assign optimal resources based on the realities of the project and the environment, moving resources around to respond to variation in duration estimates and fluctuations in the workforce, to always have proper manning of the most important tasks. In *planning* we ideally stagger projects on a "virtual drum," in order to achieve "good enough" smoothing of resource loads. In *execution* we will occasionally have peak resource overloads that require management decisions for mitigation. In addition we may have more or fewer projects actually being worked in the phase that we have identified as the virtual drum than we had when we originally staggered the projects. In such cases management will have to make decisions on how to deal with the unexpected over/under-loads.

Now we come to the warnings. These are intended to foster deep understanding of the cause-and-effect behind success or failure. In the introduction I talked about the importance of following the implementation steps precisely, and that omissions can be devastating to ultimate success. Here are a few more pitfalls of which to be mindful:

WARNING! The instructions in this book are straightforward, but straightforward does not equate with easy.

The fact is that a productive implementation takes a lot of hard work from committed employees. Both understanding and action is required from all levels of employees, from the front-line worker to top management. But of even greater importance is it should be understood that all TOC applications involve paradigm shifts. It is my opinion that a successful Critical Chain implementation requires both more and larger paradigm shifts than any other application. Some of them are huge, challenging what have been longstanding and widely-accepted norms, appearing at first radical, perhaps even frightening. In fact, nearly every step in Level 5 of the S&T tree represents a major paradigm shift. It is not easy to bring people through even one small paradigm shift, let alone several huge ones. Very simply put, IT IS EASIER TO FAIL. Or put in a better way, it is easier to run into an obstacle, blame the consultant, the implementation team, the CEO, or the software, and return to your comfort zone—your old ways of managing (late) projects.

This endeavor is not for the faint of heart. Are you serious about becoming a high performance projects organization? IT CAN BE DONE. The results can be spectacular, but the organization must successfully make the journey through the paradigm shifts. I have done my best to point out these shifts, and to clearly detail the logic behind their necessity in order to make this journey a bit smoother.

WARNING! This manual is not a substitute for qualified expert help.

Nothing can take the place of experience. As important as each step is in achieving maximum success, the instructions by necessity are still to some degree generic. There are nuances in some environments that only an experienced implementer will fully understand. Once again, if you are serious about becoming a High Performance Projects Organization, you will do what it takes to protect yourself against failure. I strongly recommend you use a qualified, experienced Critical Chain consultant to guide you through the implementation. A Theory of Constraints International Certification Organization (TOCICO) expert certified in Project Management is your best bet, but there are other very good people out there. But don't be afraid to check references, even with the certified experts.

When I was an engineer at Boeing, I read a paper written by one of Boeing's Senior Technical Fellows. I will paraphrase here something he wrote that I felt was very profound:

The fastest way to do anything is to do it right the first time, no matter how long it takes. Likewise, the cheapest way to do anything is to do it right the first time, no matter how much it costs.

Going it on your own in order to save a few bucks is a risk you probably shouldn't take. There are at least two important reasons you should consider before taking this risk. The first is that your implementation, if it succeeds at all, could yield lesser results, take much longer and cost much more than doing it right the first time. The second reason is that once you have disappointed the organization, starting over with an expert might be much harder than it otherwise would be.

WARNING! Follow the instructions Step-By-Step.

With a couple of exceptions which will be highlighted, it is essential that each step be completed, with satisfactory answers for the "three questions," before moving to the next step. The general rule is to have only one active implementation step for a given department or responsible manager at any given time. There is a great temptation to jump ahead and to work on multiple things at the same time. You are excited. Things are going well. You have a bit of spare time, another step looks easy. Starting another step won't hurt. DON'T DO IT. The cause and effect logic of the

S&T tree is meticulously crafted and tested. If you jump ahead it will cause bad multitasking, consume your capacity, and produce less than optimum results.

Per department or responsible manager: One. Step. At. A. Time. Period!

WARNING! Beware of sub-optimization.

Critical Chain is a systemic solution for project management. Implementing Critical Chain in an area, shop, department, etc. that is only part of a larger system, where the actual pacing mechanism should exist outside the implementing area, is sub-optimization. Sub-optimizing a non-constraining part of a system yields zero real (bottom line) improvement for the company as a whole, and can even be damaging. There is also a possibility of launching wars within a company when you create unnecessary demand for upstream departments or suppliers, or flood downstream links in the chain. Always try to implement at a systemic level, where the entire larger project is fully included.

WARNING! Software is usually necessary for a successful and robust implementation, but it is decidedly not sufficient.

I'll say it in stronger terms: Software may be helpful, but is *never* the solution. In fact, I contend software is the *smallest factor* in a successful implementation. That may seem odd coming from someone connected with Critical Chain software, but it is my sincere belief. The paradigm shifts and their resulting changes in behavior are by far bigger factors. It could even be said that the first few foundational steps in the S&T tree, which have nothing to do with the Critical Chain application *per se*, are the most important steps of all. If we don't properly implement the first steps, the best software in the world will not be sufficient to create a successful and sustainable implementation.

Unfortunately, people frequently tend to rely too much on software. Sometimes we will even refer to Critical Chain itself or the Critical Chain implementation by the name of the software being used. This is the height of absurdity! It would be nice if we could push a button and watch all our problems disappear. It would be nice to push a button and let the software do all the hard work. But this is pure fantasy. Software makes things much easier and gives us quick, actionable information, but it cannot lead us through paradigm shifts or make difficult management decisions for us. I urge you to always strive to keep the role of the software in proper perspective.

WARNING! The properly implemented organization is likely to become very, very productive—so productive that it can severely damage or kill your company.

What?! How can this be? There are at least two things of which we need to be aware. First, as with all TOC-based applications, a proper implementation will reveal a huge amount of hidden capacity – capacity that has been consumed in the past with; among other things, resource hoarding, Parkinson's law, bad multi-tasking, rework, and recovery from poor management decisions. You will begin to execute projects much faster than before, which is great, but you could be creating unoccupied "gaps" in your Master schedule. I think it's criminal to reward great work with layoffs, so realize that not taking action to secure additional work, by increasing sales activities in anticipation of success, will lead to slack times for resources. This means everyone will not be busy all the time.[5] This is actually a good thing, but often we live in a culture that thinks free time is a waste, and that waste needs to be cut. Workers are savvy enough to see this situation developing from a mile away. In anticipation, they frequently slow down to match their pace to the available workload so as not to reveal the actual improvement in capacity.

In truth, a reasonable amount of free time is not a waste but an asset—it is protective capacity, an essential mechanism to allow organizational flexibility/adaptability. The real problem is the existing paradigm that says "keep everyone busy" is a cornerstone of effective management practice.

---

[5] I highly recommend that all managers read the book "Slack: Getting Past Burnout, Busywork, and the Myth of Total Efficiency" by Tom Demarco, Broadway Books, 2001. I can't say it better than Mr. Demarco.

So why not instead get ready to sell more projects? As soon as you are convinced of the results—as soon as you begin to see your once-hidden capacity—you should be ready to offer something enticing to the market.[6]

And last but not least, not a warning, but a:

SURPRISE for Critical Chain Consultants! An implementation is itself a project!

Please allow me to make an assumption. Many people are reading this book because they hope or expect to implement or work in the implementation of Critical Chain in an organization.

Companies or single consultants may be implementing single projects. Larger consultancies may be implementing multi-projects. This means that they are Projects organizations, and EVERYTHING IN THIS BOOK APPLIES TO THEM FIRST, including optimal assignment of resources, PREPARATIONS, and project staggering. So be sure you read this book in its entirety first, and then go back to page 1 and implement yourself. Don't become the constraint in your own customer's success.

[6] The Theory of Constraints has knowledge and applications for creating Marketing Offers, and every Strategy & Tactics tree has a place for generating increased sales. But that's another book for which we have neither the time nor the space here.

Chapter 1 -The Team

This book is about detailing the necessary and sufficient actions to complete a Critical Chain Project Management implementation that has the highest possible probability of dramatic, sustainable success. Every chapter, every subject, every step is designed to do just that, and every compromise leads to the possibility of a less than optimal solution. When it comes to "hard actions," i.e. the actual S&T steps, compromise is to be avoided at all costs. However for the "soft" components of the implementation, such as the assembly of an implementation team and surrounding support system, we have a bit more latitude, although we should still strive to be as close to the ideal as possible. What is described in this chapter is the Ideal Implementation Team and Support System.

Do not skimp on appointing the right people to the team. Proper project management and the delivery of projects on time, on budget, and with their desired scope intact are worth untold millions of dollars to most companies.

Millions. Of dollars. Get it? Do not skimp.

One U.S. Navy client once told us that the entire implementation, including consulting and software, paid for itself in the first two months of the implementation with reductions in overtime alone. Now *that's* ROI! Done right, these kinds of results are possible. Therefore put your most forward-thinking people on the team. Do not try to cut costs or corners. Allow the team to have all the time it needs to create the best implementation possible. The more people involved the better. There is really nothing else more important when so much is at stake.

Nevertheless, some companies literally might not have the resources required to build the ideal team. Many smaller companies have this limitation. Still, if you fall into this category, strive to do your best and the payoff will be great. Although the ideal is not always attainable, substantial results can still be attained. Let me begin, then, with a foundational recommendation for all implementations, regardless of the size and implementation capacity of the organization.

The U.S. Marine Corps Albany, Georgia Maintenance Center implementation could be considered the most effective and longest-lasting implementation of Critical Chain in the world. Since first implemented in 2001, Albany has continued to lead the country's maintenance depots in productivity, continually growing in results. I was fortunate to have been part of that implementation as the senior consultant on-site. A couple of years after we had finished our work I went back for a visit. I was astonished at how the success of the implementation was continuing to grow over time.

The Maintenance Center's commanding officer during most of the implementation was Colonel Steve Foreman. He was a highly respected, visionary and effective leader. I thought it would be good to ask his opinion on why the implementation had been such a smashing, sustainable success and how we could replicate it elsewhere. His answer was quite enlightening:

"First, you need a strong leader. That's me. Next, you need the head of the production department (or engineering, distribution etc. as the case may be) to be fully committed. And finally, you have to have a representative of the workers actively supporting it." Colonel Foreman's head of production, civilian Darren Jones, had enthusiastically and openly supported the implementation. As of this writing he remains at the Maintenance Center in a key leadership role. Union steward Bert Black, who later became the union president, was the representative of the workers. Bert correctly saw that the Theory of Constraints was the future of job security for the workers at the Maintenance Center. As the results began to appear, the new orders poured in, and in a world of cost cutting and downsizing, Albany bucked the trend by hiring hundreds of new workers (by that time increasing overall production capacity by roughly 400%)!

By the way, it must be noted that of all the implementations I have ever seen, *Albany is the most successful because, in my opinion, they deviated the least from the instructions in this book.*

It's hard to argue with Colonel Foreman's advice. Seeing is believing. So this is the foundational recommendation. If you want your implementation to be truly effective and sustainable, be sure that you have on board and openly supporting:

1) The top manager of the organization
2) The head of operations, engineering, etc. for the implementing group
3) A representative of the workers, who can speak their language, and make the case for Critical Chain as being just as much in their best interest as it is in the interests of the owners, customers and management's.

All of these people should have a good understanding of Critical Chain theory, the Projects Strategy & Tactics tree, and this book. Once this foundation is in place, you can begin putting together the implementation team. The most important team member to be selected is the manager who is responsible for overseeing a successful implementation:

The CCPM Implementation Manager

This *must* be a high level employee – don't compromise on this – a middle manager does not have the authority necessary to get people to follow the rules and see the big picture. More importantly, he or she usually doesn't have the ear of the executive team. The CCPM Implementation Manager should be viewed by the entire organization as having the authority of top management. Therefore the higher in the organization this person is the better.

It is strongly recommended that the CCPM Implementation Manager have no responsibilities other than making CCPM work in the organization. It should be his or her full time job. He or she could be called Director of Continuous Improvement, or something similar. In fact, I recommend this become a permanent, upper management position. Just because the consultants go home and conditions improve does not mean that the implementation can be ignored. It must be nurtured, capitalized upon, and sustained.

The CCPM Implementation Manager should have a deep understanding of the Theory of Constraints in general and Critical Chain Project Management in particular. He or she should know the Projects S&T tree and this book by heart, and with great clarity of understanding. This is necessary in order to properly ask and test the answers to the "3 questions," which will be explained in Chapter 4. But the CCPM

Implementation Manager should be just that – a manager. He or she needs a team of qualified and committed people to do the "in the trenches" work of implementation—training, coaching, ensuring quality, reporting results, etc.

Local implementers

This is the team that makes the implementation happen. They are the "in the trenches" people of the previous section. Sometimes called "implementation leaders," they are present before and after the implementation, and at every step in the implementation itself, planning, scheduling training, attending review meetings, being the first line of information for users, etc. They deeply study Critical Chain theory, the Projects S&T tree and other TOC subjects and hopefully become the future "TOC Experts" for the company. Often they will seek TOCICO certification in Project Management and other disciplines. The selection of these people should be from among those who seem to understand CCPM naturally (they "get it"), and exhibit real enthusiasm about TOC and continuous improvement. Although not a strict requirement, it is also helpful if at least some of these team members have a strong background in project management and scheduling fundamentals. Other team members with experience in facilitation, have good people/coaching skills and have developed and implemented new business processes in the past can also be invaluable to the team. Properly trained, the implementation leaders can become the foundation for an ever-flourishing company, continually returning many times their cost in value to the company.

For larger companies, implementation team members are dedicated to the transition effort on a full-time basis and must therefore be relieved of day-to-day responsibilities until they are no longer needed full-time to carry the load of the implementation. This role can be contracted out to experienced implementation specialists if desired.

The Full-Kit Manager

The Full-Kit Manager is a crucial member of the team, and is a permanent position. He or she is important enough to merit a detailed job description (find it in Chapter 12), but for now suffice it to say that he or she is high enough in the organization to have the authority to give a green light or red light to projects, based on the availability of the full kit (*documented* completion of all full-kit requirements.)

The Auditor

The Auditor is an "outside" member of the team. If at all possible, having an un-biased set of eyes to audit your implementation can be very beneficial. The Auditor provides this set of eyes. He or she is a TOC and CCPM expert with a proven track record. In dealing with the day-to-day of implementation, it is possible to get so wrapped up in details that the local implementation team can sometimes "fail to see the forest for the trees." Being an expert on the Projects S&T tree, and with several implementations to his or her name, the Auditor reviews the implementation periodically to ensure everything is being done correctly. This person should not be from your organization, and should not be the consultant if you use one. It could be perhaps a higher-level employee of the consulting company, or better yet an independent TOC expert. In such a case the Auditor is not influenced by the politics of either organization and can also review the performance of the consultant.

An audit can take a day or two and should be held every couple of months until the implementation is an established success. If the team uses the recommended reporting process, it should be easy to prepare for each audit. A good auditor can be a bit expensive, but the value he or she brings more than pays for itself, as the audit is likely to help you maximize the potential of your implementation and avoid stagnation, where instead of making forward progress, the implementation loses momentum.

The Consultant

Also an "outside" member of the team, the Consultant is the voice of experience and the frequent visitor and overseer of the implementation. More detail about the consultant's job is found in Chapter 3, "Guidelines," under the heading "The proper role of consultants."

Departmental Champions

Last but not least, the implementation is greatly strengthened by what I call "departmental champions." They are not implementers per se, since they don't do much in terms of actual implementation work. But they are the local "cheerleaders" for the implementation. They are where the rubber of the theory meets the road of reality. Having local champions means that it isn't "them" who want "us" to implement Critical Chain. Having local champions means "we" are implementing.

Champions should be selected, by department, based on their interest and enthusiasm for CCPM. They should have sufficient training in the S&T tree to be able to answer the question "why?" when the people within their departments are inquiring why they are being asked to change their paradigms (long established policies, procedures and practices). Local champions should receive recognition and support from all other members of the implementation team, as well as from top management.

IMPORTANT! Here's a tip that will save a lot of time and frustration later in the implementation. The implementation team and local champions should create a "map" of the organization, specifying exactly who, by name, requires what training and must attend what sessions. A schedule for their training should be produced. Top management should already be trained by this point (see Chapter 3 – Guidelines), so this is the training for everyone else. Shop floor personnel and lower-level personnel in other functions should receive at least a short, 2-4 hour overview of Critical Chain. Lead persons and first line supervisors should get more extensive training, middle managers, the implementation team and champions should get expert-level training, etc. Training sessions should be mandatory and excuses should not be accepted. In order to move the implementation quickly, there should be no more than one alternate/remedial session at each level.[7]

Finally, I highly recommend a "presence" of top management in ALL official sessions, including all training sessions. Relatively speaking, this is a small thing, but it can have a highly significant positive influence on the implementation. Believe it or not, even when top management is completely committed to embracing CCPM in the company, a common complaint from line employees is, "Management is not serious about this." This is because despite management's best intentions, all the new conditions will not be in place overnight. Therefore, there will be plenty of evidence to prove that "nothing is different." If there is not enough visible and tangible evidence of management's commitment to seeing the change through, it will be easy for employees to conclude that it's just another "flavor of the month."

In addition, when senior management is expected at the sessions, subordinates are much less likely to find excuses not to attend. The executives' role can be as little as spending the first five minutes welcoming people to the sessions, with an affirmation of top management support for Critical Chain and the implementation team and then returning at the end of the session to field questions and to make sure people don't duck out as soon as they leave.

Do everything you can to put together the team described here. Are you serious about becoming reliable? Then (while there are sure to be a lot of urgent distractions,) very few things are likely to be as important as this. If there truly are more important issues relating to the long term growth and stability of the organization, then you are well advised to make sure and deal with those items before turning your attention to launching the implementation. Once the implementation has been officially launched, then you want to make it clear to those involved and affected that NOTHING is more important. A (strong and properly resourced) good team is a key element of a successful Critical Chain implementation.

---

[7] It is a good idea to schedule on-going sessions for new employees, etc. But the purpose of the mandatory initial sessions is to not allow "foot draggers" to slow the implementation.

CHECKLIST:

1) Top manager visibly and strongly supporting CCPM?
2) Head of engineering/development/operations/production fully and publically committed to implementing CCPM?
3) Representative of the workers promoting CCPM as the employees' best interest?
4) High-level Implementation Manager selected and relieved of other responsibilities?
5) Implementation team appointed and trained?
6) Auditor selected and reserved?
7) Departmental champions selected and trained?
8) Training "map" of organization created and distributed?
9) Presence of top management in all sessions?

Chapter 2 – Strategy & Tactics

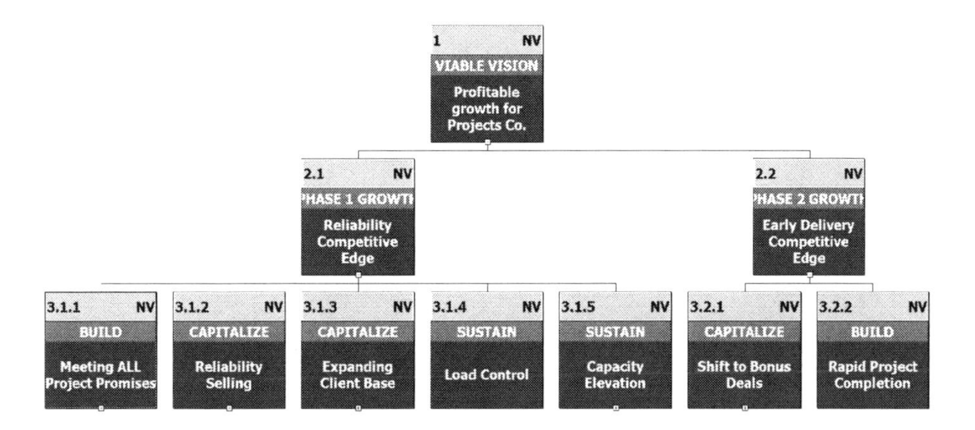

The graphic above, resembling an organization chart, is actually the index of the top three levels of what is called the Projects Strategy & Tactics tree. It was developed, along with S&T trees for various other environments, by Dr. Eli Goldratt in the first decade of this century. (The little "plusses" on the bottom of some of the Level 3 boxes indicates additional levels below.)

The Projects S&T tree in its entirety details all the necessary and sufficient steps to completely transform a projects-related company into an "ever-flourishing" company, "continuously and significantly increasing value to stakeholders - employees, clients and shareholders." It is really about turning an entire company into one that thinks differently and performs continually better and better. As Eli used to say, "Even the sky is not the limit." Dr. Goldratt calls the quest for the ever-flourishing company a "Viable Vision."

Notice that the full S&T tree deals with many things: building successive decisive competitive edges, an enhanced sales process, etc. Notice also that as the branches of the trees go "down", they direct their focus on three areas: Build, Capitalize, and Sustain. In other words, building the mechanism to become a highly reliable deliverer of projects; capitalizing on this newfound power; and last but certainly not least, sustaining the ever-flourishing company over time.

Before you are tempted to dismiss the S&T trees as a bunch of technical words and rules on a cute little chart – something to be skimmed over and forgotten, let me offer you this: In November 2008 Dr. Goldratt and I were invited to a Ministry of Land, Infrastructure and Transportation conference in Sapporo, Japan by the Japanese government. After the conference Eli and I were walking down the street, and I asked him a question about the S&T trees. His response to me was startling at first. He stopped, turned to me and said, "I consider the S&T trees to be the most important thing I have ever invented." This threw me for a loop, and at first I couldn't believe it. This was the man who had introduced us to Drum-Buffer-Rope Production Management, Critical Chain Project Management, TOC Distribution & Replenishment, the TOC Thinking Processes, and the Five Focusing Steps. I knew there was value in the S&T trees, but could it really be the most important thing he ever invented?

Later I heard Eli repeat this statement many times, often in front of large crowds of people. And with every passing day I have come to believe more and more that Eli was correct. The S&T trees are the missing "grail" for creating success in our implementations. They are the Rosetta Stone, the key to understanding. They establish the *what, why* and *how* of every implementation step. Maybe the most important of these is the *why*. The S&T tree delivers the answer. Why this way? Why not another? Using assumptions, or statements about reality, Goldratt uses the S&T tree to carefully lay out the logic of the *why*. Because it represents the cumulative experience of many years, many experts and many different implementations – some more successful than others – its credibility is well founded. This is the reason that by following the S&T tree as closely as possible, a company maximizes its probability of a highly-successful Critical Chain implementation. This roadmap therefore significantly reduces the implementation risk for any organization embarking on this journey. Without following such a roadmap, the organization is left to rely too much on the experience and intuition of the individuals entrusted with the implementation.

Another definition of the S&T tree is that it is a vehicle to transfer a vast amount of information to a large number of people in a quick, clear, and effective manner. Even the casual glance at the Generic S&T tree reveals that there is a lot of information being covered—information on nearly every aspect of building, capitalizing upon, and sustaining the ever-flourishing company. I highly recommend that companies think about having S&T trees tailored for their organization, and that they consider embarking on a "Viable Vision" for their company.

For the purposes of this book, we will not be looking at the entire S&T tree. We will be focusing on the box identified as 3.11.1 – Meeting ALL Project Promises, and its level 4 and level 5 entities below. As you will see, even what is contained in 3.11.1 alone requires an entire book to fully describe and explain the "what's, why's and how's" of implementing Critical Chain Project Management. And even though we are only dealing with 3.11.1 and below, please don't forget the warning in the "How To Use This Book" section, about dealing with the soaring productivity in your organization as a result of implementing Critical Chain.

Structure

We turn now to the basic structure of the S&T tree—how to understand the connections, the hierarchy among the individual entities, and the logical process involved in each entity. The best place to start is to consider the "conventional wisdom" or paradigm of what the definitions of "strategy" and "tactics" actually are. Goldratt's claim is that the current paradigm is wrong, and that a correct understanding opens up possibilities far greater than the conventional wisdom allows.

Dr. Goldratt said that for many years he read every book on strategy and tactics that he could get his hands on, and no real coherent standard emerged between them. The definitions were never quite satisfying, and the view of strategy and tactics was somewhat nebulous, mysterious, and abstract. But they all essentially followed a paradigm, that strategy was something that happened at the top of an organization and tactics was something that happened "down in the trenches" in an attempt to accomplish the strategy. A graphical view of this paradigm looks something like this:

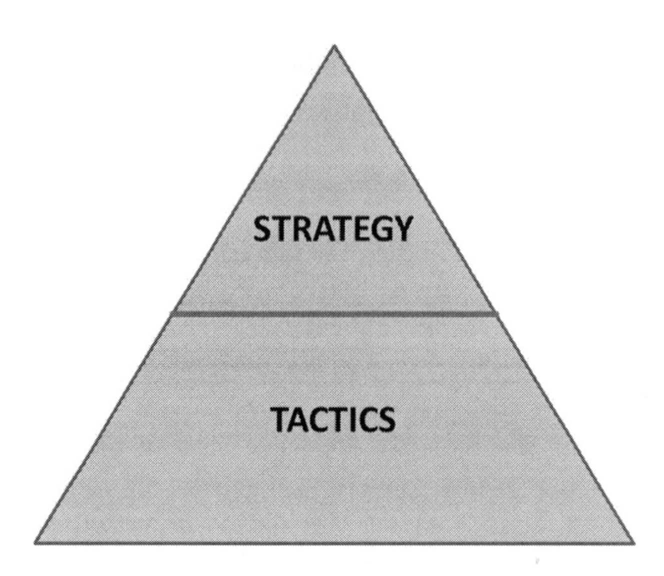

Figure 1. Traditional view of strategy & tactics

Goldratt saw that the real problem with this paradigm is the line between the two concepts. How does one take this thing "strategy," formulated in the board room by highly paid executives, and turn it into practical tactics that can be executed by line workers on the shop floor? Even the transfer between the executive suite and the next level of management is problematic. How can workers many levels below ever understand how the things they are being asked to do on a daily basis relate to the distant corporate strategy? Even more mysterious, how can they know if what they are doing is contributing to the success of the company? For this reason, corporate strategy tends to end up being little more than a poster on the wall in the supervisor's office. In the end, management's full strategy is rarely achieved, and what is accomplished seems to take forever. Frustrated managers try changing tactics, sometimes entirely abandoning earlier ones. Workers feel nothing is well thought out or permanent. In their eyes, top management is incompetent. In the eyes of top management, workers are stupid and lazy. Neither view is correct. But the lack of a coherent and integrated definition of strategy and tactics makes it very difficult for different levels of an organization be on the same page.

Goldratt concluded that a new paradigm was required—one where strategy and tactics are paired together at every level of the organization.

This is the real genius of the S&T tree. The *what* (strategy) for each step at each level is clear to everyone, as is the *how* (tactics) to achieve the strategy.[8] In this way the strategy has direct, accessible substance, and the connection to the tactics is clear and understandable. The tactic is the cause that enables the desired effect (the strategy).

Here is a view of Goldratt's paradigm of strategy and tactics:

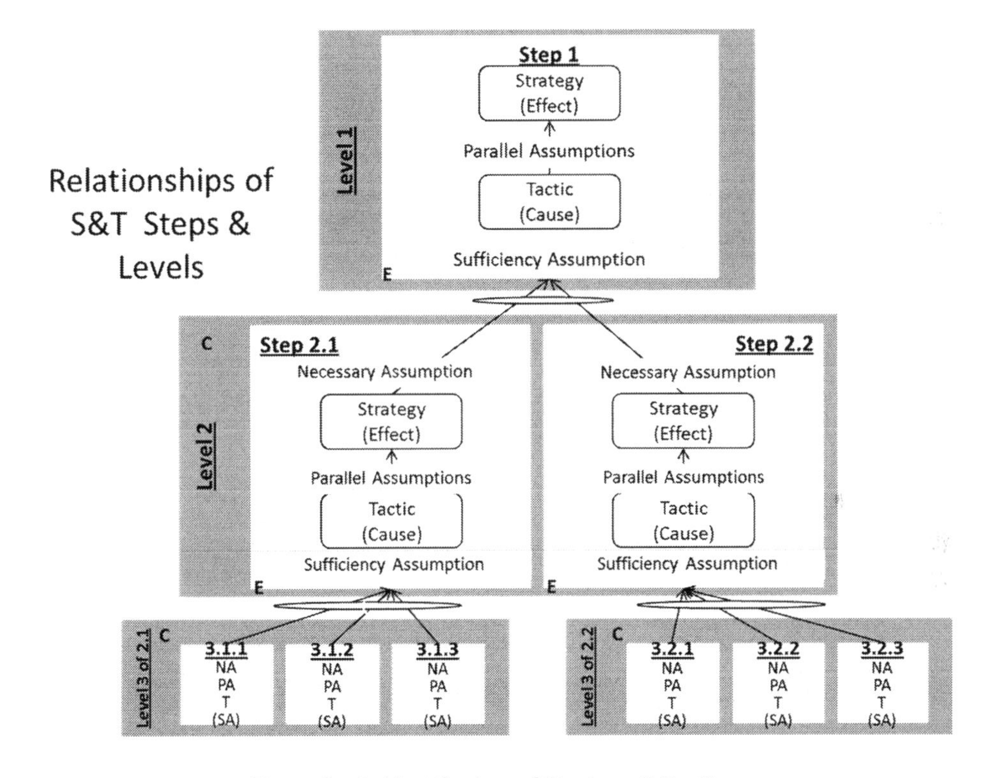

Figure 2. Goldratt's view of Strategy & Tactics

Notice that in the diagram above, there are some other components to the S&T tree entities: assumptions. In an S&T tree there are three types of assumptions: Necessary, Parallel, and Sufficiency. We

[8] Common sense reveals this to be obvious. Even simple actions in everyday life have paired "what" and "how" elements. What: I want to walk without danger of tripping and falling. How: I tie my shoelaces. What: I want to find something in the dark. How: I turn on a flashlight. Why we have strayed from common sense and separated strategy from tactics in the business world is an interesting question.

will consider each of these types of assumptions in a moment. But is important first to stress that assumptions are just that—assumptions. Assumptions are not facts. Assumptions can be either valid or invalid.

For example, I assume that in the spring, if I cultivate my home garden in the Seattle area, plant seeds in early April, provide water and fertilizer and keep the weeds away, come fall I will have a nice vegetable harvest, as I have had in past years. It is a reasonable assumption—something in which I can put a good deal of confidence.

But my assumption could turn out to be wrong. My seed could be old and fail to germinate. A late-season freeze could wipe out my seedlings. Insects could destroy my plants or their fruit. Therefore it behooves me to check (test or challenge) my assumptions. What are the dates on my seed packets? What is the average and the latest historical frost dates for my area? Do I chance planting earlier? What insect-mitigation methods do I have in place?

In addition, assumptions that are valid for me in my environment may not be valid in other environments. I might be able to plant certain crops in Columbus, South Carolina in early April, but could I plant the same in Fairbanks, Alaska at the same time? I might be able to successfully grow oranges in Florida, but could I assume the same success in western Montana?

Assumptions should be scrutinized, tested and challenged. When we can have confidence in our assumptions we are able to build good tactics and realize our strategies. Even the assumptions in the S&T trees need to be challenged and tested. Some customization might be required for a particular environment. This could mean different assumptions and different tactics. Just be sure that any customization to the S&T tree is led by a qualified person, proficient in the TOC Thinking Processes.

Now let's take a look at the different types of assumptions in the S&T tree. What is their definition and importance in the S&T structure?

Anatomy of the S&T	
Necessary Assumption (NA)	The "Why" of the Step. The reason that the higher level S&T step cannot be implemented unless a change is made. In other words, it describes the necessity for an action to be taken.
Strategy (S)	The "What" of the Step. The objective – the intended outcome – of the S&T step. When the strategy is achieved, the need highlighted by the necessary assumption is met.
Parallel Assumptions (PA)	The "Why" of the Tactic. The conditions which exist in reality leading us to a specific course of action that would achieve the strategy; forms the logical connection between the tactic and the strategy, explaining why the tactic is the course of action that leads to the attainment of the strategy.
Tactic (T)	The "How" of the Step. What needs to be done in order to achieve the strategy. In a well written S&T step, the tactic is obvious once the parallel assumptions are read.
Sufficiency Assumption (SA)	The "Why" of the next level. Explains the need to provide another level of detail to the step. If we don't pay attention to it, the likelihood of taking the right actions is significantly diminished. (SA is not in lowest level of the S&T)

Figure 3. Anatomy of the S&T tree

Necessary Assumption: The Necessary Assumption tells us the reason we need a strategy—why creating this particular strategy is necessary to our success. Once we know why we need a strategy, we can create one from a foundation of understanding. The Necessary Assumption exists at the top of every S&T entity with the exception of the very top entity—the overall company strategy. We can all *assume* that every company needs a top-level strategy, can't we?

Parallel Assumption: One might be able to think of many different tactics to achieve a particular strategy. But are all tactics created equal? Does it not stand to reason that some tactics will be more effective than others? If so, how do we know we have chosen the best, most effective tactics to secure our strategy? The parallel assumptions lead us there. All S&T entities contain Parallel Assumptions.

Parallel Assumptions are statements about how we perceive reality. There are usually, but not always at least two Parallel Assumptions in an S&T entity. It is possible that a single assumption is so powerful that it can lead to a tactic without an additional assumption. There also can be many more than two parallel assumptions. There should be as many as are required to choose a powerful tactic.

To illustrate, let's consider a particular strategy: let's say, for example, we want our company to "Dominate the portable music player market." In most cases, at least one assumption is "positive." For example, "There is a huge worldwide market for portable music players." Yes, this is an assumption, and in the early 2000's it was a pretty good one.

Likewise, in most cases at least one assumption is "negative." "Many companies are capable of producing music players at low cost." This assumption is a statement of perceived reality. The challenge is to choose a tactic that will capture market share in the face of extreme cost competition—and make a nice profit doing it. Maybe another assumption should be added to help lead us to that tactic: "There is currently no easy way to get all the songs a particular customer might want."

These assumptions, existing in parallel like lines on school paper, led Apple to a winning tactic: "We create a comprehensive music store from which our device can effortlessly download music." Of course, Apple likely did not use an S&T tree. But then again, not every company has a Steve Jobs, whom it appears, followed the process in his head. Unless you are Steve Jobs, this is not recommended.

If you do have a Steve Jobs on your team, you might not need an S&T tree. However, if you want a structured method that can help you develop, clarify and communicate your strategy effectively, then an S&T tree is a great way to go.

Sufficiency Assumption: The Sufficiency Assumption indicates that there is not yet enough information to fully achieve the strategy, even if the chosen tactic is solid. "We create a comprehensive music store..." is a great tactic, but how do we create it? Is it available to everyone? Only Apple users? Does it work with multiple operating systems? How much should one pay for a song? The list goes on. To fully reach the desired strategy, more detail is necessary. This will necessitate an additional, lower level of the S&T tree. It is possible it will take more than one additional level.

There also could be more than one Sufficiency Assumption. The Sufficiency Assumption is also used as a warning that without further detail, actions taken can lead to negative effects. A Sufficiency Assumption in the above case could be something like, "It is important to the record labels that their music be protected from theft."

How many levels should an S&T tree have? The guideline here is that once you have enough detail that you reach the point where people know exactly what to do without a written instruction, (or, if the written instruction already exists in another document, such as an operating procedure) additional levels are not necessary. Therefore when this point is reached, the lowest level entities do not have Sufficiency Assumptions.

At present, most of the S&T trees I have seen have five levels. This is the case with the generic Projects S&T tree. Some S&T trees may have four levels. On rare occasions an S&T tree may only have three levels. None ever has less than three. I have also seen a few trees with six levels, and even one with seven. You create as many levels as are required to ensure clarity, then you stop and get to work.

You should be able to see now why the S&T tree is so powerful, and how it enables communication and clarity at all levels of the organization. With a good S&T tree, even the lowest level employee knows why he or she is being asked to do what they are doing, and how it contributes to the success of the company as a whole. They are even able to follow the tree upwards and understand the thinking and strategy at the very top of the company. For the workers, it is possible to conclude that maybe the executives aren't incompetent after all. At the same time the executives learn that when the workers understand how they contribute, they somehow magically become smarter, harder workers.

In order to get a sense of what is entailed in a Critical Chain implementation, and why this book is so long, I thought I would show you a graphical representation of the S&T steps for implementing Critical Chain. Each of these entities have a corresponding chapter, and each contain strategy, tactics, and assumptions. Even so, before we get started, we need to lay a bit more groundwork.

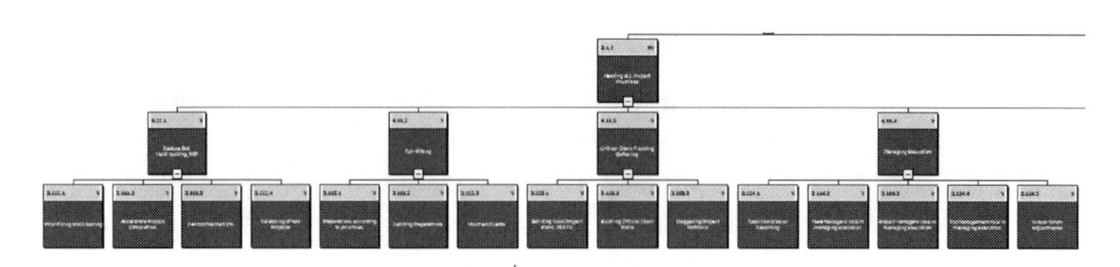

Figure 4. The twenty S&T entities for implementing Critical Chain. All are covered in this book.

Chapter 3 – Guidelines

As you prepare to embark on implementing Critical Chain in your projects environment, it is beneficial to set some guidelines—or ground rules—that will help keep the implementation on track and keep focus in the proper place. These guidelines come as a result of years of experience as a consultant and implementer of Critical Chain and other Theory of Constraints applications. As with the more specific rules which accompany each implementation step, they are designed to increase flow by removing obstacles and minimizing rework. Like the lane-dividing lines on a road, staying within these guidelines affords the smoothest, fastest way to the goal. Drift outside them, and you risk driving into a ditch, or worse, into oncoming traffic.

Ensure Top Management Buy-In

When considering the ideal Critical Chain implementation, the first and foremost priority is the unequivocal buy-in and public support of top management. In fact, the truly ideal implementation would be initiated and directed by top management itself. As one might expect there are many important reasons for top management buy-in. Foremost among these are:

1) Critical Chain will affect productivity and profitability.
2) Critical Chain involves paradigm shifts that will conflict with conventional measurements and conventional wisdom, including corporate measurements related to efficiency, labor hours collection, and measures related to methodologies such as Earned Value Management.
3) Critical Chain employs a "start projects as late as possible" philosophy, which can create alarm among executives, sister divisions, and especially customers who do not understand the benefits of such action.

4) Top management is responsible for the prioritization and cessation of work on existing projects during the initial phase of implementation, and re-prioritization of projects thereafter.
5) Under Critical Chain rules, in many environments only top management can authorize the insertion of a new, emergent project into the pipeline. They also have the responsibility to provide guidance on which affected project(s) on which to stop work (temporarily or permanently). In addition, top management must also ensure how and when to report to customers changes in delivery dates for projects that may result from a re-balancing of the portfolio.

It should be clear why top management buy-in is so important. If it cannot be acquired, you should give serious thought as to whether to attempt a CCPM implementation at all. At a very minimum, if you are in a large company, you should get authorization from the top level management of your division, i.e. someone with P&L responsibility and decision-making authority. If even this is not possible, yet you still want to proceed, be prepared for the pain, frustration and possibly failure that might result.

NOTE: Despite the above warning, it is also true that we owe much of our insights and experience to many courageous individuals who implemented CCPM in less than ideal leadership conditions.

Finally, top management buy-in and understanding is important to the ideal Critical Chain implementation from an S&T knowledge perspective.

Responsibility of each level of management to train immediate subordinates on S&T tree

Although some training can be led by the implementation team, the ideal way to transfer the basic S&T knowledge throughout the company and help ensure sustainability for the future is for each level of management to first obtain knowledge of the S&T tree for themselves and then train their immediate subordinates. For example, the consultant could train the top management team, and then each top manager would train their direct reports, with this pattern repeating itself all the way down through the organization's hierarchy. The consultant and/or implementation leaders could be present to oversee and contribute if necessary. Imagine how this process would foster understanding throughout the organization at every level.

The proper role of consultants

As was stated previously, a company is wise to use an experienced and certified CCPM consultant when implementing Critical Chain. However, since the Theory of Constraints differs fundamentally from the conventional wisdom on many key points, we should consider carefully what a CCPM consultant actually is, and what he or she should be expected to do.

Whereas the traditional consultant is seen as someone who performs analysis, creates reports and renders advice to the client organization, the CCPM consultant is generally much more results-oriented. Anything less than *actual* significant shortening of projects, improved on-time delivery and more company throughput is unacceptable. He or she is largely uninterested in providing only "take it or leave it, just pay me a lot" advice.

Usually, although not always, the business model of the TOC consultant differs from that of a traditional consultant, in that a time and materials approach is set aside for a firm, fixed fee or gain-share approach, where a smaller base amount is charged with large bonuses paid based on meeting agreed-upon performance objectives. This is again an indicator that the TOC consultant is results-oriented.

Being results-oriented means the consultant will spend considerable time on-site, transacting with everyone from the CEO or President of the company to the task manager on the shop floor. If the job description of the consultant is not carefully defined, this can lead to some negative effects that will not serve the company well in the long run. For several reasons, it is sometimes too easy for the consultant to find themselves in a compromising position regarding the best way to serve the interest of the client. We find that very often organizations today have so "leaned out" their organizations that they have very little discretionary capacity in place to do very important work related to getting better or healthier as an organization. At the same time, consultants by our very nature are knowledgeable on a wide array of topics and situations and live for the opportunity to apply ourselves to solving our client's problems. Here are a few pitfalls that can arise in an engagement if both the client and the consultant do not take steps to prevent them from happening:

1) The client becomes too dependent on the consultant.
2) The client expects the consultant to do the *work* in the implementation, thus preventing the transferal of knowledge and skill to the company.
3) The consultant, who is often considered (rightly or wrongly) to possess superior knowledge of business in general, gets sidetracked by organizational issues that are either not central to, or are not even related to the successful implementation of the Strategy and Tactics elements.

Despite the otherwise good intentions of all involved, we believe that the proper role for the TOC consultant in the implementation is to be a guide – a referee – a sky diving instructor, if you will — explaining the rules, showing the way, pointing out dangers – but not actually doing the work himself. It is your company, your future, and *your implementation*.

Although the consultant may train top management and other key people, he or she should not be expected to train everyone. Since it is your implementation, once understanding is achieved, your own company personnel (management hierarchy – not a support function) can and should train others. In this way the high degree of ownership in, as well as knowledge of all the important elements of the implementation necessary to sustain the results long term, is permanently transferred from the consultant to your company.

Other initiatives

Most organizations are characterized by a history of a continuous stream of corporate initiatives. HR initiatives, process improvement initiatives, IT initiatives, government compliance initiatives, health and safety initiatives, quality initiatives – the list goes on and on. Unfortunately very often companies try to implement too many initiatives at the same time, leading to company-wide bad multi-tasking and disappointing results. Like those listed above, a Critical Chain implementation is a major initiative, especially considering the numerous and profound nature of the paradigm shifts involved. Most important among those paradigm shifts is the realization that an organization's effectiveness is greatly improved when it has the discipline in place to focus attention and resources on a few things, as opposed to diffusing or diluting attention and resources across too many things. Therefore when intending to implement CCPM, the organization should think very carefully about what to do with other initiatives which are in progress or planned to start in the near future.

Which initiatives are most important? How many are too many? Certainly you would not want to attempt a CCPM implementation at the same time as say, installing a new ERP system (unless it is to use CCPM to ensure that the roll-out of your ERP implementation is executed quickly). A choice must be made: is it more important for me to have a new business system or to start bringing in our projects on time and under budget? If on-time projects are more important, the ERP implementation should be suspended and put on hold until CCPM is fully in place. If the ERP system is more important, CCPM should wait. However when making the decision, consider each initiative's impact on the bottom line and corporate goals of the company. Politics may also often be involved. If your goal is to be the company with the best integrated computer system and don't have to worry about reputation and profit, put CCPM on the shelf. But if your goal is happy customers, increased sales and profitability, and a harmonious work environment, well, you know what to do!

Chapter 4 – The Three Questions

We are nearly to the place where we will begin to examine the Projects Company S&T tree and discover the roadmap for the effective Critical Chain implementation. First however, I want to share three basic questions that will help to supercharge the S&T tree implementation process, making it come alive and providing greater results and effectiveness for your organization. The questions are simple, but their importance cannot be minimized.

The Three Questions were developed over multiple TOC implementations and are my offering to the TOC community not only to enhance the Projects Company S&T tree, but all S&T trees and all implementations. There is an implementation pitfall that both the company and the consulting team can sometimes fall victim to if we are not careful – the temptation to "check the box," go through the motions or take credit for progress on a step too early – in order to keep on schedule or to give people a warm, happy feeling. The Three Questions have the ability to make the S&T trees more relevant and powerful by creating a standards criteria and feedback loop.

The Three Questions provide the following benefits:

- They put the proper amount of focus on the S&T tree
- They demand the definition of criteria for knowing that the step has been implemented correctly, and the organization can confidently proceed to the next step
- They help manage the continued effectiveness of the implementation over time
- They are appropriate to be asked by all levels of management, spreading knowledge and awareness of the success of the implementation throughout the organization
- They help prepare the organization for an audit
- They help to validate or invalidate assumptions about the S&T steps as written

Practical examples will be given for how to use The Three Questions in many of the following chapters. In addition, a section is included at the end of each Level 5 S&T steps for you to fill in your answers to the questions. Let's now take a quick look at them:

1) Was the step correctly implemented?

On the surface, this may seem like an odd question to ask as you believe you are completing an S&T step, but without it, the answer can be subjective and you can get inconsistent results if implementing in multiple areas of the organization. The obvious need to be filled for this question to be powerful is clear criteria for knowing the meaning of both "completed" and "correctly." In some cases I have tried to provide the criteria here. In others, the organization may need to discuss and debate what exactly constitutes completion and what constitutes a correctly implemented step. This discussion can be very valuable to the organization in helping to construct solid thinking processes.

2) Were the expected effects realized?

Like the first question, the criteria for answering whether or not the expected effects were realized will generate valuable discussion in your organization. First, we need to clearly define what the expected effects of our tactic should be. This is not as easy as you might think. As you will see in coming chapters, a hasty answer to this question can often be wrong.

Specifically, people tend to "get ahead of themselves," assuming the step should produce effects that by itself it would never logically be able to produce. In other words, the answer to this question may point out what is necessary, but not what is sufficient. Consider the following statement and the related expected result. Do you see a problem with it?

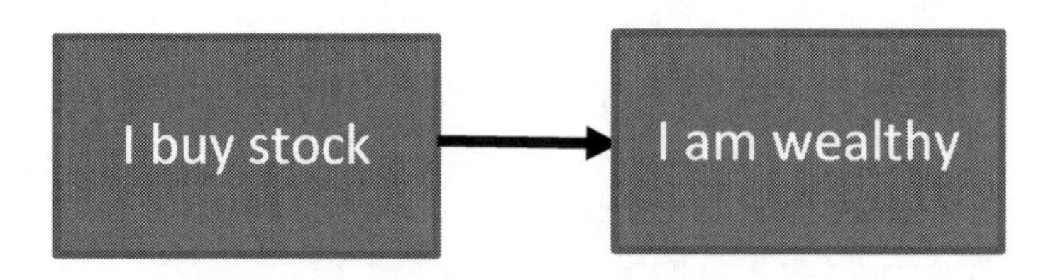

Figure 5. Example of insufficient cause

If my strategy is to achieve personal wealth, and my chosen tactic is to invest in the stock market, then it is necessary that I buy stock. But the mere purchase of stock does not mean I will be rich. The purchase of stock by itself is not sufficient—many other intermediate things must occur. The real result of my purchase of stock is much more immediate, practical and far less exciting:

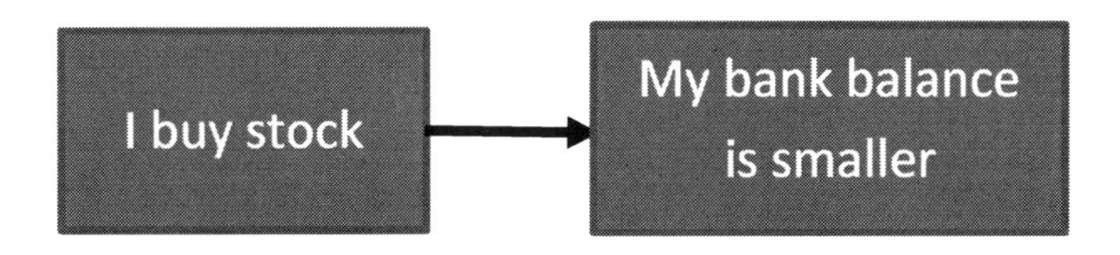

Figure 6. Example of immediate effect

It is important that we clearly specify the actual, immediate result of our action. Only in this way can we best determine the next set of strategy and tactics to continue the path to our goal. An excellent CCPM-specific example of this will be shown as we discuss 5.111.1 in Chapter 7.

Even after validating the logic, although one would expect that in most cases if a step has been implemented correctly the answer to Question 2 should be "yes," experience proves this is not the case 100% of the time in all implementations. The fact that we implemented the step correctly is not an absolute guarantee of effects. The possible answers to Question 2 include "yes," "no," and "we're not sure." So what does it mean if we are certain we have implemented the step correctly but we are not seeing the expected effects? In such a case (which is somewhat rare) we have what Dr. Goldratt called a "mystery," because there is usually no quickly apparent reason the effects are not forthcoming.

When we have a "mystery," we have to go back to the S&T tree and study it carefully. We need to check the "facts of life" about our environment that are written in the "Necessary Assumptions," "Parallel Assumptions," and "Sufficiency Assumptions," in the earlier entities of the S&T tree. Usually if we find an invalid assumption it will be in the Parallel Assumptions. If we have an invalid assumption we can choose the wrong tactic for achieving our Strategy.

By closely reviewing our assumptions and finding one or more that are in reality not valid, we are better equipped to construct a modified or different tactic that will bring us the effects we expect. In extremely rare cases we might even find we have chosen the wrong Strategy. But experience shows that in most cases projects organizations are not all that dis-similar and the basic strategies and tactics laid out in the S&T tree will deliver the expected results.

3) What mechanism has been put in place to ensure compliance with the step over time?

The S&T steps, or more precisely the intended effects of the steps, should not be thought of as one-time events, but rather they should create conditions that can be sustained over time. For example, if we reduce the load on the system we need to make sure it does not grow again to the point our results deteriorate. Therefore we need one or more "mechanisms" to help us ensure our implementation stays healthy in the years to come.

Very often the "mechanism" will involve measurements or metrics – something that alerts us when results are in danger of deterioration. I have endeavored to suggest adequate new measures and metrics with each Level 5 S&T entity. But this brings up another question. What do we do about our current measurements and metrics? Keep them? Throw them all out? Keep some and toss others?

In general, there may be some current measurements and resultant behaviors that are obviously contradictory to the basic concepts of CCPM. In such a case these measurements and behaviors should be changed (or eliminated) immediately. A good CCPM consultant can help you recognize these types of contradictory measurements before you begin the implementation, and keep you from having to learn the hard way. Other than these you should keep your current measurements and metrics until it becomes apparent they are obsolete under the new system or in some way cause behaviors that hamper results. If a current measurement conflicts in any way with the strategies and tactics we have designed for success, we need to take action as soon as possible.

Armed with *The Three Questions* we have a great (much improved) chance of successfully implementing our tactics and achieving our strategies. In the next chapter we will begin our analysis of the Projects Company S&T tree, entity 3.1.1 and below. If you are implementing the entire S&T tree, congratulations! If not, I encourage you to review the entire tree and see what you might be "leaving on the table." Either way, please review the warnings once again in the "How To Use This Book" section, and take a close look at entities 3.1.2 "Reliability Selling" and 3.1.5 "Capacity Elevation" in the S&T tree, which are not specifically covered in this book.

Chapter 5 – 3.1.1 Meeting ALL Project Promises

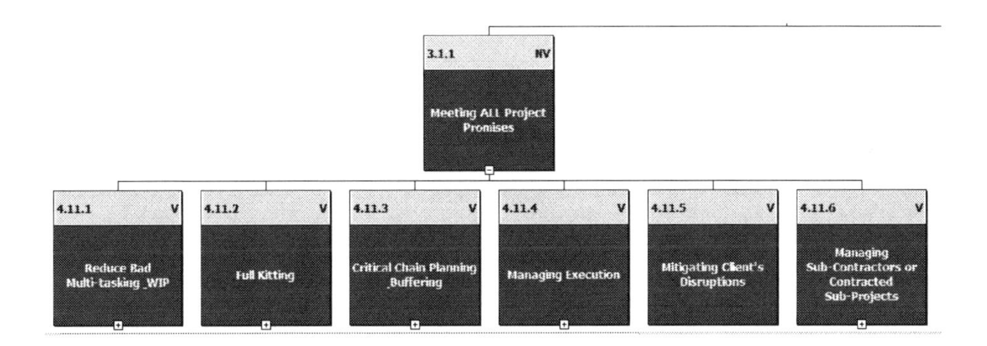

Now it's time to clearly specify what we are trying to accomplish by bringing Critical Chain to the organization. For that, let's turn to the Projects Strategy & Tactics Tree. In the figure above we see the titles for the entities in Levels 3.1.1 and Level 4 below. Beneath these, not shown, are the Level 5 entities. There are seventeen Level 5 entities associated with 3.1.1. These are the true action steps, and we will discuss each of them in detail.

We see that the title of 3.1.1 is "Meeting ALL Project Promises." Notice that the word "ALL" is in capital letters. This is to emphasize that if we are to be a truly effective projects organization, we must consistently meet a number of criteria. First, it means that we will (almost) always deliver projects on time.[9]

But our promises to our customers to deliver on time are not our only promises. Next, it means that we will (almost) always complete our projects within the original budgeted cost. Indeed, we will deliver on time,

---

[9] In a world where 40% project due date performance is often considered excellent, we are instead talking about delivering significantly more than 90% of our projects on time, from now on, forever.

but we will not ask the customer for more money, nor will we eat into our profit margin.

Finally, it means that we will meet our promises of delivering the full intended original specifications of the project. We will not cut corners, defer features or settle for inferior specs. Meeting ALL project promises means meeting all the promises of schedule, cost, and scope. At this point you might be thinking, "It's impossible." But I assure you it is not. It has been done. It is *being done* today in project organizations around the world. And you can do it too.

But simply waking up one morning and proclaiming, "From now on we will be on schedule, on budget, and delivering full specifications," is obviously insufficient. It stands to reason that in order to achieve such outstanding performance, we need a plan. We require a set of strategies and tactics to get there. The S&T tree contains these strategies and tactics, along with the assumptions used to create them. 3.11.1, "Meeting ALL Project Promises," begins by defining why a strategy is needed:

Necessary Assumptions

Not meeting promises (especially when hefty penalties are involved)[10] may bring a company to its knees.

Whenever one or more of the three promises of projects are not met (schedule, cost, scope) there can be serious ramifications both within the company and with your customers. Some examples: if extensive overtime is used in order to deliver on time, the company's profit can be slashed or even eliminated. Holding costs under control can cause schedules to be missed and specifications to be trimmed.

[10] A Viable Vision for a projects-oriented company typically includes a marketing offer that ensures "hefty" monetary penalties will be paid to the customer for late delivery of projects. But hefty penalties can also exist in a more traditional sense, e.g. the delay in progress payments for large projects such as aircraft or ships; or when a penalty is paid for late delivery by contract. One way or another, monetary consequences can be associated with late deliveries for most project organizations.

Completing all specifications can consume profit and cause schedule delays. Whoever the customer is, whether internal or external, he will suffer, along with your organization's reputation. Repeated too often, this could drive the customer elsewhere. On the other hand, delivering on time, on budget, and with full specifications will usually have the opposite effect. Old customers send more projects your way and new customers line up at the door seeking your services. With such extreme outcomes at stake, we need a strategy to help us keep ALL of our promises.

Strategy

The Company has very high due-date performance without compromising on the content or on the budget. In the multi-project arena, very high due-date performance is defined as delivering well over 95% on (or before) the original promised due-dates, while in cases of late delivery the delay is much smaller than the prevailing delays in the industry.

Here the official strategy of the company becomes "We successfully do the impossible, day after day, month after month, year after year." How? Remember, the strategy is the "what." The tactic is the "how." We're going to need a powerful tactic or set of tactics to get us there, and the parallel assumptions help guide us to the tactics.

For most projects-oriented companies, being on time to internal or external customer promise is of great importance. But before we take a look at the Parallel Assumptions, let's consider the organization that may claim that on-time delivery is not a necessary condition to their success, such as an R&D organization or software developer.

Granted, some projects companies do not have to commit a completion date to a customer. But does that mean that a completion date is irrelevant? Does it mean that reliability is not important, even if the promise is to yourself?

For example, you want to release software version 3.0 in the fourth quarter of the year. You want to do it within your budget and you don't want to defer features to later releases if at all possible. I claim that reliability is very important, even if this reliability is only to yourself. You should always have a specific date that you are working to, or "fourth quarter" will somehow magically become December 31st – and you might find yourself competing against other "fourth quarter" projects, paying a lot of overtime the last two weeks in December and deferring features anyway.

It is always beneficial to work to specific dates – and to meet those dates – even when you don't have to commit dates to a customer. Furthermore as you will see, when projects are correctly staggered, you will see if "fourth quarter" is actually achievable, or if it is a pipe dream. It might even be possible to complete early in the fourth quarter, or even sooner, thus bringing in throughput in greater amounts and earlier than you had hoped for.

This said, all S&T trees are customizable.[11] I have actually changed this specific strategy at a software developer (and therefore we had different tactics), because we could better satisfy the market's significant need in a way that none of our significant competitors could (refer to Projects S&T tree Level 1). But if you are one of these organizations, think very carefully about this and discuss it with your consulting team before deciding to change strategy or tactics in the generic Projects S&T tree.

Parallel Assumptions

Most compromises on content or budget stem from the pressure to meet the promised due date. Critical Chain Project Management (CCPM) brings most multi-project environments to high due-date performance without compromising content or budget.

Here we have an assumption that Critical Chain Project Management can "do the impossible" in most multi-project environments. Remember, assumptions are not facts. Assumptions require challenging, testing, and evidence that they are valid. In the case of Critical Chain, vast experience has shown us that CCPM brings most multi-project

[11] Changes to the S&T tree should only be done by qualified and experienced writers of S&T trees, who are certified in the TOC Thinking Processes.

environments to high due-date performance without compromising content or budget.

We believe this is a valid assumption for a great majority of organizations, regardless of their industry, products or markets. Nevertheless it is still possible that in some environments Critical Chain might not be the best answer. Remember, assumptions are there to be challenged.

Tactics

The Company implements Critical Chain Project Management (CCPM) culture and procedures.

In order to achieve the strategy of meeting ALL project promises, the tactic most often chosen by companies knowledgeable about TOC is the implementation of Critical Chain Project Management. But notice the tactic does not say merely to implement CCPM procedures. It also says to implement CCPM *culture.* Changing culture is much harder than changing procedures, because changing culture is not technical endeavor – it involves changing paradigms and mindsets. So although the sentence describing the tactic is short and to the point, the effort required to execute the tactic is not trivial.

The S&T tree itself is a primary tool for educating the organization, helping to foster the required culture change. This is a major reason the S&T tree should be considered not just as "nice to have" coaching, but actual company policy combined with measurements and corrective action for violation of the procedures. If your organization is serious about outstanding project performance, it should be serious about the Strategy & Tactics tree.

Sufficiency Assumptions

To ensure an outstanding start of a major initiative it is vital that the first substantial actions will result in immediate substantial benefits.

In terms of creating successful, sustainable CCPM implementations, this assumption may be the most important one in the entire S&T tree. I cannot stress this enough. If I had the ability I would put it on a movie marquee with flashing lights and mount it in the executive offices. This is the key to successfully implementing *any* management initiative, not just CCPM.

The early steps in the Projects S&T tree are designed to deliver immediate substantial benefits to the organization. That is why the actual Critical Chain steps and implementation of software come much later in the process. Getting immediate, substantial benefits is the primary reason we say that if you are implementing any other major initiative you should not try to implement CCPM at the same time. It is also an important reason for the optimal assignment of resources (5.111.2 – see Chapter 8 – "Accelerate Project Completion.") If your resources are spread too thin, everything will take much longer than it has to, and if your initiatives take too long, interest and enthusiasm can be lost – people start asking why they are not seeing benefits – and all initiatives start looking like "flavor of the month" failures.

Resources should not be exhausted and time should not be wasted. Focus!

The sufficiency assumption is warning us that our knowledge is insufficient. It is telling us we need more detailed information. Yes, most people will agree we need immediate, substantial benefits. But how do we get them? To find out, we need to dive down one or more levels. They should give us more instructions on the "how to" of getting these benefits.

Section Two: Reducing the Load on the System

Flow is the number one consideration

Chapter 6 – 4.11.1 Reduce Bad Multi-tasking, WIP

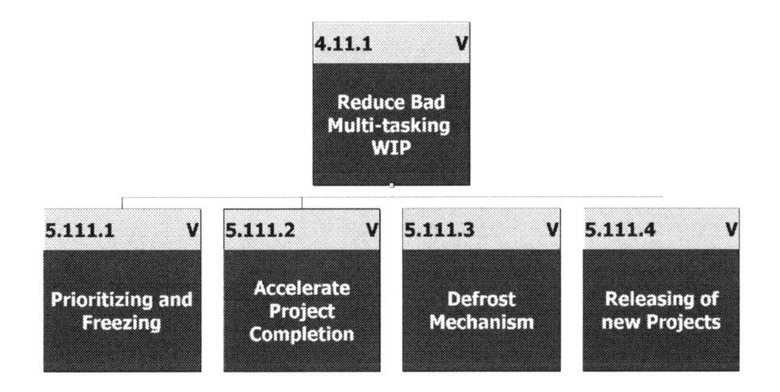

The Sufficiency Assumption at 3.1.1 informed us that more information was needed for us to execute our tactic of implementing CCPM culture and procedures. This created the need for a fourth level of the S&T tree. Under 3.1.1 there are six Level 4 entities. All are necessary for us achieve our strategy of meeting ALL our promises, and only if we have completed all six can we have good assurance that we can achieve this strategy.

The proper way to implement an S&T tree is to work from the bottom, left to right, (almost) always one step at a time. Therefore we will now move down the left side of the tree and turn our attention to entity 4.11.1. I have titled this section of the book "Reducing the Load on the System," because in practical terms, that is what entities 4.11.1 and below do. Reducing the load on the system not only makes projects flow much faster (similar to the way reducing cars on the highway at rush hour increases flow), it dramatically reduces many opportunities to fall into the great time-destructive behavior known as multi-tasking.

The actual title of 4.11.1 is "Reduce Bad Multitasking, WIP." This is a clue that in addition to reducing bad multi-tasking we will reduce the load on the system by removing Work-In-Progress. We will discuss this in detail in coming chapters. But the titles of the entities don't really tell us much, and in fact they are not intended to. To get to the real meat of the matter, we have to "open up" the entity to see the strategy, tactics and assumptions.

Let's take a close look at 4.11.1. Once again, we "pair" strategy and tactics at every level of the S&T tree in order to ensure clear communication, providing all employees understanding not only for the "what" of the step, but also the "how." In this way, in the bigger picture, everyone can also understand the "why," and understand how what they do contributes to the company's success. First, we need to consider why we need a strategy for reducing bad multi-tasking and WIP.

Necessary Assumptions

When too many projects are executed simultaneously, many resources will find themselves under pressure to work on more than one task - bad multi-tasking is unavoidable. Prolific BAD multi-tasking significantly prolongs each project's lead-time.

We will begin by discussing bad multi-tasking, and then we will talk about reducing the number of projects in the system. When approaching the subject of multi-tasking, it is important to note that the word itself can be somewhat misleading. True multi-tasking would literally be doing two things at the same time. An example would be sweeping the floor while talking on the phone. With multiple-core CPUs, computers nowadays can literally multi-task.

But this is not so true with people. In the business world, there is very little true multi-tasking (two or more tasks in active execution by the same person in parallel). Rather, there is an abundance of *task switching*, or stopping work on one task prior to completion in order to start working on another task. The worker is only doing one task at a time, but he or she is stopping and starting tasks rather than completing them. This task switching is what we refer to in TOC as "bad" multi-tasking. For our purposes, please consider the two terms interchangeable.

NOTE: It should be understood clearly that task switching is the number one enemy of completing projects on time.[12] Stopping and starting tasks is by far the worst thing you can do if you want your organization to be reliable; to meet ALL its promises. This is why the first steps in the Projects S&T tree are concerned with eliminating multi-tasking.[13] Without reducing task switching, the CCPM steps to follow will be of little value. If want your organization to improve and you do nothing else, strive to eliminate task switching. This step alone will provide more benefit than any other action.

A quick search of the web will reveal that a large and growing number of studies confirm the negative effects of multi-tasking. There are also simple games to see for yourself how bad it is. I have included one here. You can try it on your own or in your group. It only takes a few minutes and participants only need a blank sheet of paper and one or more pencils or pens. You will also need someone to keep time, using a clock or stopwatch.

First, draw two vertical lines on the paper, creating three columns. Each column will contain a series of tasks. Label the columns "Letters," "Numbers," and "Symbols." In the left column, participants will write the letters A through Z. In the center column, they will write the numbers 1 through 26. Finally, in the right column, they will alternate the following symbols: square, circle, and triangle.

---

[12] This includes stopping tasks for things such as status meetings or other types of meetings.

[13] In certain specific cases, task switching may not always be considered "bad." For example, if someone working on a time-consuming, low-priority (green) task is asked to stop work in order to help on an emergent high-priority (red) task, and therefore the overall organization is benefited, it is permissible. However if these are not rare exceptions, i.e. if poor planning is constantly shifting priorities and people are frequently being asked to stop work to move to another task, the benefit to the organization is lost—"bad" multi-tasking has set in and all projects will begin to move slower and slower and due dates will be missed. The key is that this action must demonstrably help recover significant buffer. Keep these occurrences to an absolute minimum, reserving them for true emergencies.

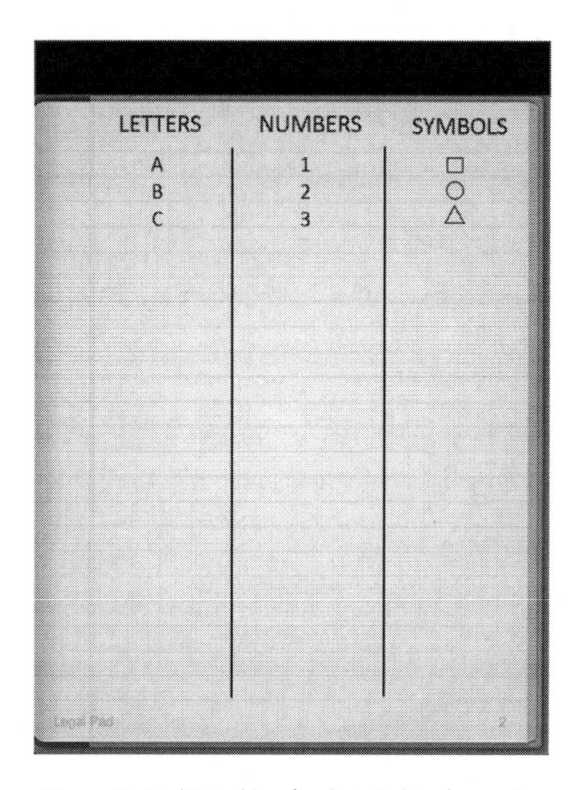

Figure 7. Multi-tasking (task switching) exercise

Do the exercise twice—once with, and once without multi-tasking. To simulate multi-tasking you will only write three characters at a time in each column, and then switch to the next column and write three characters there. Therefore you will write A, B, C in the left column, move to the center column and write 1, 2, 3, then move to the right column and draw a square, circle, and triangle. Now return again to the left column to write D, E, F, etc. until you have 26 characters in each column. Time yourself or your group, noting the duration from beginning to completion, or in the case of a group, the range of durations.

Next, do the exercise again without multi-tasking. In the left column write the letters A through Z, then switch to the center column and write the numbers 1-26, and finally fill in the entire right column with alternating symbols. Once again record your durations. Compare the first duration or range of durations with the second. You will see that you or your group were significantly faster the second time around, when you were not multi-tasking. Of course in the real world there are rarely as few as three tasks ready for work, but it's good enough to give you the idea.

Yet even this does not adequately portray the damage done by stopping and starting tasks. In the real world there is set-up and set-down time associated with moving from task to task. One might have to put drawings and tools away and get new ones for the next task. Then one has to remember where they left off, and orientate themselves again on the approach to each task. Therefore you might try the exercise a third time (or just substitute this for the first run), simulating the addition of set-up and set-down times for each instance of task switching. Use different color pens for each column, and put them away after each use into a drawer or pencil case, making you have to retrieve them again for the next use. Time yourself again and compare your time to the second (non-multi-tasking) run of the exercise. You'll be amazed at the difference. And remember, in the real world you will likely have many more than three tasks ready to execute.

Also, in the real world you will likely have more than one project active at any given time. You may have several, or scores, or even hundreds of active projects, depending on your products and work environment. Can you imagine switching not only between three columns, but between three columns on each of many pieces of paper? Oh yes, and don't forget status meetings!

An additional negative effect of task switching is making more errors. When you have to re-orient yourself to a task, there is a danger of making mistakes through confusion or fatigue. Errors lead to rework, and this contributes even more to delays in completing your projects.

Perhaps you can now see how devastating task switching is to completing projects on time, and why it is so very important to keep it to an absolute minimum. Therefore in order to minimize task switching in the real world, we need a strategy and one or more tactics.

Strategy

Flow is the number one consideration (the target is not how many projects the Company succeeds to start working on, rather it is how many projects are completed).

In order to limit task switching as much as possible, the strategy of the company will be to consider the holistic flow of the organization the primary judge as to whether or not take any specific action, especially one such as stopping one task to work on another. The questions will be along the lines of "What is best for the system as a whole? How many concurrent projects should be in work? What actions will move all projects more swiftly?"

Even the answers should be scrutinized again on the basis of flow. Keeping in mind that task switching is only appropriate when a true emergency allows us to consider stopping a low-priority task for the sake of a high priority task, what evidence is there that action A is better for flow than action B?

Flow is the number one consideration.

Parallel Assumptions

The statement, "the earlier we start each project, the earlier each project will be finished," is not correct for multi-project environments (not only the first elephant but also the last elephant will go through a door much faster if they go in procession). Vast experience shows that in multi-project environments, reducing the number of open projects can reduce bad multi- tasking without causing starvation of work and therefore significantly reduces the lead time of all projects - it increases the flow.

Here is another clue about what our approach will be. In addition to a deliberate, real-time avoidance of task switching, we will not start all projects as soon as their contracts are signed. Again, we will discuss this in detail in coming chapters.

But for now, to provide understanding, imagine a room full of elephants. Your first thought might be, "That's ridiculous. Elephants shouldn't be in a room, they should be in a zoo – or in the wild." I fully agree! It's best then that we get these elephants out of the room! We notice that there is only one door – barely big enough to fit an elephant through – and a couple of tiny windows. If we make loud noises in the far corner from the door, the elephants might leave the room. Let's try it. We go to the far corner, and begin to shout loudly and bash metal objects together. The elephants are startled and make for the door en masse. It's a stampede! The pachyderms push and shove for position, creating an elephant traffic jam at the door. For a while none of them seem to be making it out. Then, after several minutes, one elephant makes it through. At last! But we soon see that the traffic jam reappears, and it takes another long stretch of time before another elephant can exit through the door.

After what seems like hours, the last elephant finally makes his way out. We scratch "make loud noises" off our list of ways to get elephants out of a room. Our "start everything early" (by making loud noises) policy does not equate with early finishes of elephants through the door. We bring in shovels to clean up the mess on the floor. Finally done, we leave for the day, but we forget to close the door. When we arrive the next day we find the room is once again full of elephants!

This time we use our heads to think more clearly about the situation. We realize that when there are no "rules" about evacuating elephants from a room, every elephant looks out for his own best interest, and they all try to go first (just like project managers!) They jam the doorway and none can escape. But when we consider our *system* (one room, many elephants), we perceive that an orderly evacuation, with a primary rule of "one elephant at a time" will work much better. This time, we lead the elephants out of the room in procession, one at a time. We are surprised to see them even linking themselves together, trunk to tail, as they casually and without panic leave the room.

We are also surprised to see that even the last elephant leaves the room much faster in procession than in the "no rules" situation. The fact that we *delayed* many elephants for a short time to lead them one by one from the room actually made our system move *faster.* The delay of procession meant the room emptied in minutes rather than hours.

Our assumption is that it is the same in our organizational systems. When too many projects are in work, it's like a room full of elephants with no evacuation rules. Bad multi-tasking, or task switching, like the elephants pushing and shoving for position, becomes prolific, and everything slows down. But introducing a rule – the delay of some projects to clear the road for others, *increases the flow* of the organization as a whole.

Tactics

The Company properly controls the number of projects that are open at any given point in time.

In order to achieve our strategy of using flow as the number one consideration for completing the maximum possible number of projects in the shortest possible time, our tactic will be one of tightly controlling the number of projects that are open in our system at any given point in time – today, tomorrow, a year from now, etc. We arrive at this tactic by considering the "facts of life" outlined in the Parallel Assumptions – that starting lots of projects does not equate with completing lots of projects, and that a heavy load of projects in the organization forces bad multi-tasking, eventually causing us to fail to meet ALL our promises.

There is an implication here (experience reveals this is almost always true) that the system is currently overloaded and Work-In-Progress (WIP) should be effectively lowered. Steps will have to be taken to ensure that work is stopped on some projects, and these projects will have to be sequestered in some way. But how? We need more information.

Sufficiency Assumptions

Adjusting the amount of work is not enough. The company must also ensure that as time passes the proper amount of work will be always maintained.

Once again the Sufficiency Assumption provides us with a warning – that adjusting the amount of work by itself is insufficient. It also informs us that more detail is required for:

1) The "how to" of performing the adjustment
2) The "how to" of sequestering the projects on which we have stopped work

3) The "how to" of devising a mechanism that will make sure we do not overload our system again in the future

Therefore we need another level on the S&T Tree, Level 5. In the case of the Projects S&T tree, Level 5 usually consists of detailed "action steps;" the "how to" of implementing Critical Chain. So now, keeping to the left of the tree, we move down to where our active implementation really begins. Are you ready? Let's go.

Chapter 7 – 5.111.1 Prioritizing and Freezing

Implementation day has arrived. The entire organization is prepared and the team is in place—the CCPM Implementation Manager, the Local Implementers, the Departmental Champions, the Full Kit Manager (see Chapter 12). You have secured buy-in and visible support from top management. One or more of them are present at your kick-off meeting. Everyone in the organization, based on their level of involvement, has been introduced to and understands the S&T tree. All key people have read this book. The Sales department knows not to simply commit standard lead times for new projects.[14] Today you will take the first major step, and experience the first major paradigm shift.

Once again, the scenario above represents the ideal situation—one in which the organization is well prepared for a successful transition to, as well as the sustainability of the new way of executing projects. Please think twice before you attempt an implementation with any of these key pieces missing. The risk you are taking is that if you aren't fully prepared and can't show early and significant improvement, your implementation may die, and you might not ever have another chance to transform your company into a high-performance projects organization.

In the last chapter it was mentioned that multi-tasking, or task-switching, would be attacked in two different ways—first, by considering the holistic flow of the organization whenever making management

---

[14] Cross-functional discussions are necessary for quoting due dates during implementation. By the time implementation is complete you will have applied a new method of committing due dates with the sales department. The full discussion of this method is found in Chapter 18. However during this transition phase it will be obvious that simply quoting a standard lead time from "today" for new projects would be a major mistake.

decisions, and second by reducing the load on the system through controlling the amount of Work-In-Progress, or WIP. With 5.111.1, we take the first concrete step toward taming multi-tasking, the cornerstone of early implementation success.

Before we take the step however, we want to understand why we need a strategy.

Necessary Assumptions

Reducing the number of open projects by delaying the introduction of new projects is too slow - freezing open projects is required. It is unrealistic to expect that project managers will reach a consensus on which projects should be frozen ("I fully agree... as long as my elephant goes through the door first!").

This assumption tells us that although there may be a strong preference among some for just letting attrition account for the eventual WIP reduction, this approach is insufficient. There are a number of reasons for this, but the assumption above states a big one: *it's too slow.* The obvious question is, "too slow for what?" Here are a few answers:

1) It's too slow to properly implement the next step (5.111.2), which involves optimal assignment of resources to (the remaining) open projects.

With such an approach, the resources needed to accelerate projects won't be available for a considerable amount of time, and as a result from both an internal and external (customer) perspective, it will seem like nothing has changed.

2) It's too slow from the customer's perspective. You will still be seen as unreliable. Can you take this chance? Will your customers stick with you while you improve almost imperceptibly at first, and then very slowly?
3) Most important of all, it's too slow from your employees' perspective. Will *they* stick with and enthusiastically support an initiative that is not producing visible results? Remember the Sufficiency Assumption from 3.1.1:

To ensure an outstanding start of a major initiative it is vital that the first substantial actions will result in immediate substantial benefits.

Not only for Critical Chain, but for any initiative, enthusiasm will quickly wane and the initiative will have limited efficacy or even fail if substantial immediate benefits take too long to materialize. To rationalize the attrition approach as being "controlled and deliberate," implies an increase in the probability of success, while it actually *decreases* the probability of long term sustained success. Think of this as a conflict if you will. On the one hand, you want the benefits of less work in progress because without these benefits it will be difficult to implement the subsequent steps of the transition. On the other hand you also want to keep working on previously initiated work because the consequences of not doing so are too severe. In order to have a successful transition, it is at the same time necessary to both avoid severe consequences (risk) of each proposed step (reducing WIP) as well as secure the intended benefits of the step. The following diagram illustrates the dilemma:

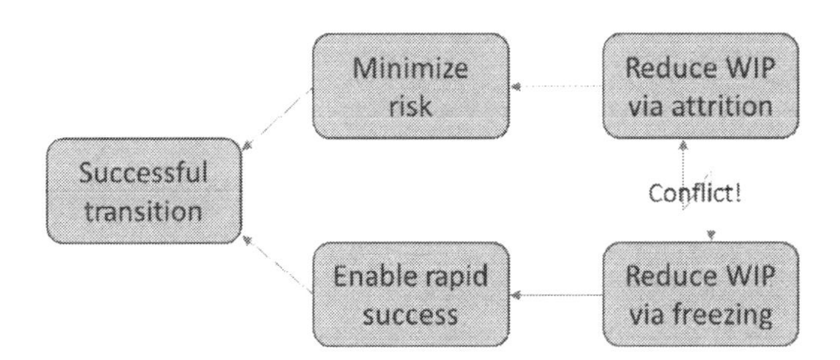

Figure 8. Work-In-Progress reduction conflict

One key to resolving this dilemma is to realize that because attrition is too slow, the need the approach addresses (minimize risk) cannot in fact be met through attrition. However it *can* be adequately addressed by pursuing the Prioritizing and Freezing step described in this chapter (the last elephant will leave the room faster than it would have otherwise). Although they feel real, the assumptions behind the perceived risk of freezing are actually not valid.[15] The Prioritizing and Freezing step

[15] Another perceived risk, that of clients being worried by the fact their projects are frozen, is addressed in step 5.112.3 (Chapter 14).

will actually minimize your risk much more effectively, while at the same time enabling rapid success.

The organization about to embark on a Critical Chain journey therefore, should clear the decks of other initiatives, take the necessary preparatory steps, and implement quickly but step by step. It is truly your best chance for success and sustainability.

Strategy

The number of open projects is quickly reduced to be more in line with better flow and throughput.

Since lowering WIP by attrition of projects is too slow, the strategy is that we will reduce the number of open projects quickly. In the tactics we will define how this will be done. The parallel assumptions will lead us to the correct tactic, one which will help not only reduce bad multi-tasking, but will also increase the flow within, as well as the overall throughput of the system.

Parallel Assumptions

In the extreme case, when there are not enough projects in execution, "Starvation" lowers the rate of projects completion. In the opposite extreme, when there are too many projects in execution, "Bad-Multi-Tasking" lowers the rate of projects completion. Between these two extremes there is a (almost) plateau. Having prolific Bad-Multi-tasking is a clear indication that a system is in the second extreme case. Reducing the load by 25% will move the system away from one extreme without the danger of reaching the other extreme. A person in charge of a cluster of projects can and should decide on their relative priorities.

We must remember that the above statements are assumptions—they are not necessarily facts. Let's take a close look at them and test their validity. If we agree that they are valid, they will help lead us to the proper tactic(s).

The first two statements relate to two extreme and unhealthy potential states of the system. The first scenario is called "Starvation," where the number of projects in work is so low that we run out of work. If there are projects in queue (a backlog of ready but not yet started projects), such a scenario means the system is running much slower than it could, and otherwise workable projects are being unnecessarily delayed. That we don't want to be in this situation seems obvious—the assumption appears to be valid. The good news is that when there is a backlog of projects the fix is easy—just put more projects into work.

Because the fix is a no-brainer, we almost never see this problem in the real world. The only reason some companies experience starvation (and this sometimes happens due to increased speed and flow after CCPM implementation)[16] is they simply don't have enough work for the capacity and resources available. This is not a logistics problem, but a sales problem.[17]

The other unhealthy situation noted here is experiencing delays in projects due to having *too much* work in progress. This scenario is harder to recognize, the inevitable bad multi-tasking has much larger implications for the organization, and the condition is more damaging to the goal of meeting all project promises.

Experience reveals that it is extremely common that organizations find themselves in this overloaded condition. Unfortunately, they frequently come to the erroneous conclusion that the problem is not that they are overloaded, but that they have a resource shortage problem. In theory, they are correct. If one could magically add the right resources at all the right places at no extra cost to the organization, then no other solution would be necessary. For most situations however, this solution amounts to wishful thinking. There usually are very good reasons why organizations do not increase resources despite obvious and chronic overloaded conditions. Chief among them are that these new resources typically come at a premium, are difficult to find, hire and train and there

---

[16] Please refer to the productivity warning in the "How To Use This Book" section.

[17] TOC deals with this situation through attractive "Market Offers" that meet a significant need of the customer in a way that no significant competitor can. Market Offers almost never compromise on price, but since they deliver something the customer really needs, they are extremely effective in drawing new business.

are no guarantees that the work will be there for long enough to support the increase.

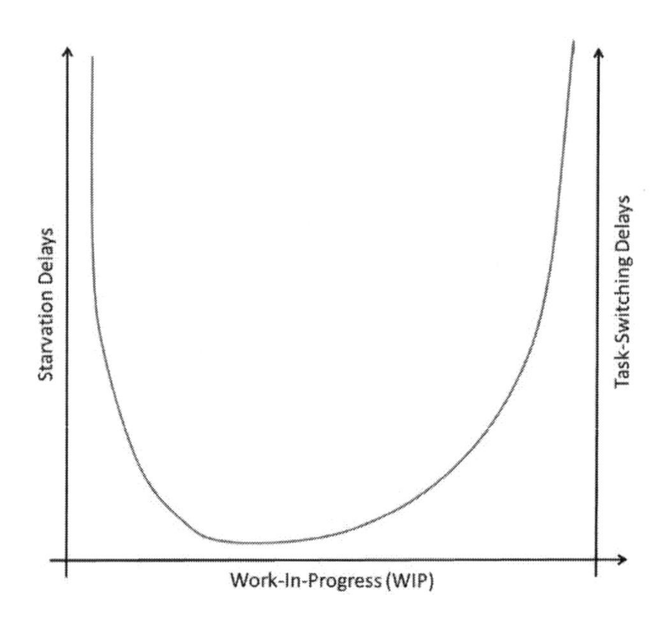

Figure 9. Notional view of delays related to load on system

Another often overlooked reason for why adding more resources to an overloaded situation doesn't always help is that new resources consume additional management bandwidth (and other overhead) to ensure that their contribution is additive to the efforts of the group. If the group is already operating with a deficit of management bandwidth (the typical case), adding new capacity without increasing the necessary overhead to go with it tends to yield less incremental improvement than hoped for, assuming there is any improvement at all. Said another way, adding more resources to a poorly managed system typically never yields the expected increase in effective capacity. To really improve the effective capacity of an organization, it is prudent to first ensure that the existing capacity is not being wasted. When this step is taken as a pre-requisite to adding more resources, it makes the process of eventually identifying, hiring and training new personnel (if still necessary) a much more highly targeted and impactful process than it otherwise would be.

The fact that projects are delayed waiting on resources tends to confirm the assumption that we don't want to be in this situation either. We don't want to starve the organization and we don't want to create stagnation of projects by chronically overloading the resources. We need to find that healthy balance between these two extreme conditions. The questions then becomes: how do we confirm that we are in an overloaded situation, and how do we find that healthy balance that allows the existing organization to complete the most projects in the shortest time?

To find the extent that key personnel are being robbed of their productive capacity as a result of too much forced multi-tasking, you should measure and analyze it. Take a week or two to observe a representative sample of the organization. Debrief with the workers after your observations and encourage (be prepared for) some candid feedback. Ask them how often they are asked to stop working on one task and switch to another. If it turns out that observations and discussion reveal multi-tasking is not a significant concern for your organization (not a daily occurrence), congratulations! You are among the small minority of organizations that have managed to escape this energy-draining condition. If, on the other hand, you are like most organizations, you can then be confident the prolific task switching you discover is a result of being overloaded. Measuring multi-tasking again after implementation will confirm the amount of hidden capacity you were wasting due to task switching.

We want to be on the "plateau" between the two extremes described above. If we experience prolific bad multi-tasking, it is confirmation that we are not on the plateau, but in the second extreme condition. Sometimes Parallel Assumptions include assumptions about the direction of the solution. That is the case here. The assumption is if we reduce the load on the system by 25% (e.g. remove 25% of projects from work), we will move away from the overloaded condition and in the direction of the plateau. It appears that it would certainly move us in that direction—it clearly would reduce the load on the system to some degree and move us away from the most extreme delays caused by the overload. But the question remains: "is 25% enough to reach the plateau and maximize flow?"

In this case the assumption sounds good, but is not as strong. Since the creation of the Projects S&T tree, growing experience is revealing that

25% should only be thought of as a bare *minimum* for reduction in load, not a target. You should test this assumption in your own organization. Don't be surprised to find that although 25% would be helpful, it may not be enough to bring you to peak performance. Too much bad multi-tasking could still exist. More information on choosing the proper amount of load reduction is discussed below. Now let's define the tactic to cover a wide range of situations.

Tactics

The top manager in-charge of all projects, after consulting with his subordinates, determines the prioritization of projects and instructs to freeze (cease activities on) enough* of the lowest priority projects.

***"Enough" means: responsible for at least 25% of the load. The proper actions are taken to ensure full adherence to the freezing decision.**

Freeze can be a powerful, even scary word. Perhaps when you hear the word, you think of the ancient glacial ice of Antarctica. Perhaps, if you are a project manager and your project is on the "freeze" list, you think your project will be incased in ice and buried in a white wasteland forever, never to be seen again. For some reason, people often associate ice with permanence.

With CCPM, we should have no such fear. For us, "freeze" is synonymous with "pause." Some of the frozen projects will actually be paused for a relatively short time. All projects will be "defrosted" within a period of time where "not only the first elephant but also the last elephant will go through a door much faster if they go in procession." In other words, even the project defrosted last will complete earlier than it would have had we not implemented CCPM.

Still, putting on hold projects already in work is a tremendous paradigm shift, and it is sometimes difficult to get people to really comprehend that in order to finish some projects earlier we need to put them on hold. Yet this is the foundational step for the rest of the implementation. We cannot afford to compromise—all that is to follow is built upon this step. S&T tree education can help people see the "big picture" and help guide them through the paradigm shift.

There are two important factors that block the adoption or acceptance of the Prioritizing and Freezing step in organizations. One is the inability to appreciate that by temporarily suspending work on some projects, we are able to gain enough speed to offset the delay caused to those suspended projects. The other is the failure to recognize that Freezing is already taking place – now – on a micro-level in the current environment, but it is hurting rather than helping the projects to complete faster. What the above tactic is accomplishing is to gather all these tiny delays that are hidden in the day-to-day operations and concentrating them in big visible blocks of time. A direct benefit of this move is that the time to actually work on each project is dramatically compressed – due to the elimination of the interruptions.

It helps to think about this in the following manner:

Every time you stop working on one task to start working on another, you are freezing the first task. It will not be defrosted until you are able to return to it, re-acclimate yourself, and get the task set up once more for work.

If yours is like most organizations and you are trapped in chronic bad multi-tasking, you are freezing and defrosting tasks (and thereby projects) *constantly*. You are Boreas, Greek god of winter. Like Midas and his golden touch, everything you touch turns to ice. You freeze so much and so often that projects are *always* delayed—to the point that long projects and late deliveries are considered a fact of life, the nature of the business.

Worse yet, this undisciplined, random freezing is done without an objective and without a plan—unlike the Prioritizing and Freezing step proposed by the S&T tree. Through this one-time action of removing load from the system, existing projects are less heavily delayed overall, and with the eternal spring of CCPM, delays in future projects are drastically reduced. That is why this first step is also the most important.

Before we continue our discussion on reducing load, let's return to the tactics and take a moment to consider who should be making the decisions surrounding the Prioritizing and Freezing step. The tactic clearly states it should be the top manager in charge of all projects. Please note that this is *not* referring to a portfolio manager that a larger company might employ, but rather someone who has the final P&L responsibility for all projects. The reason for this is the implications involved with the implementation of CCPM in general and freezing projects in particular. The actions you will take will have major effects on your organization and your customers.

Only top management should have the authority to prioritize projects and decide which ones will be removed from work, because only they have access to all the necessary information for prioritization and freezing.[18] For example, a project that may appear to be small and unimportant to a lower-level manager may actually represent huge future business for the company if the customer is satisfied with the results you deliver. But the implications can go much further.

The financial stability and reputation of the company could be at stake. There may be milestone or progress payments involved with some projects. There may be penalties associated with changes. The viability of some projects may even have a shelf life, such as in new product development.[19] Some customers may have to be told their delivery dates are changing (you would have missed them anyway), but they can be informed up-front you are implementing improvements to be much more reliable in the future. This best comes from top management. The new dates (and those quoted for new projects thereafter), unlike your current commitments, will be reliable.

As is described in the tactics, the top manager should consult with his subordinates before making a final decision. This includes people from

---

[18] This information includes sufficient knowledge of the S&T tree and this book. If the top manager does not have this knowledge, he or she will be making a less-than-informed decision. This could result in poor results, or worse yet the failure of the implementation.

[19] Especially in new product development, some consideration should be given to "killing" projects rather than freezing them. Realizing their criteria for approving new product development was too permissive, one consumer electronics company elected to kill, rather than freeze, a large percentage of their projects. The result was an organization able to focus and increase their speed to the point of getting their new products to market significantly ahead of the competition, giving them a decisive competitive edge.

sales, production, engineering, etc. When he or she is satisfied that all relevant information is available, the decision can be made. This satisfies the question of who should direct prioritization and freezing. The next question is, how much should be frozen?

The key word here is "enough." This means not too little, and not too much. But how do we know how much that is? The one thing we are certain of is it needs to be at least 25%. But 25% is not a target, it is a minimum. If we take it as a target, and apply it to an operation that is still too overloaded, we might not get enough benefits from our implementation, or at least not get all the possible benefits (see Figure 10). The actual reduction percentage required could be much more than 25%. It could be 50%. Or 70%. Or more.

But these numbers should not alarm you. It's good news! By freezing, for example, 50% of all projects in work, we are in essence UN-freezing 100% of the remaining projects. And if you are so overloaded, so addicted to task switching that it will require a reduction of 70% of projects, you will experience tremendous improvement quickly. The chaos will fade to calm, and projects will fly through your organization faster than you ever imagined. And the last elephant will still get out the door faster.

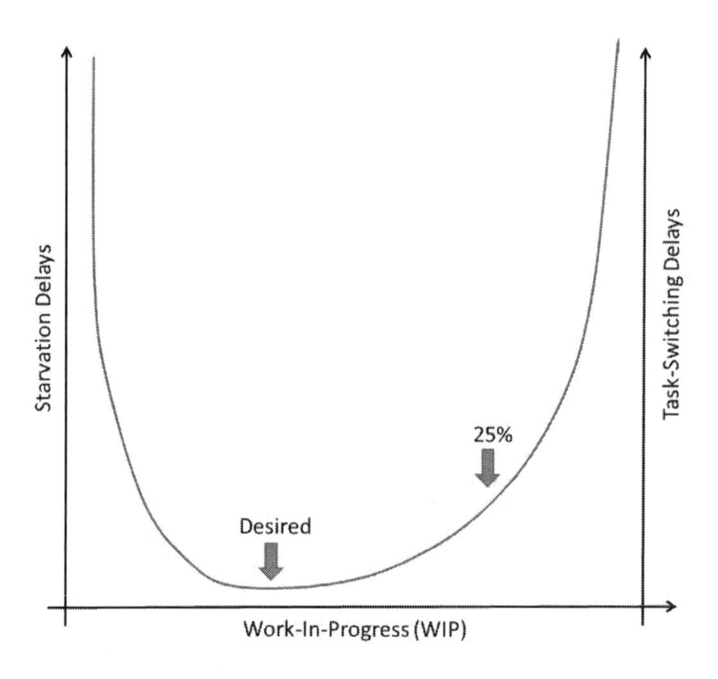

Figure 10. Possible effect of reducing WIP by 25%

We're almost there, but there are still a couple of other criteria for validating how much load to reduce on your system. Prior to beginning your implementation, it is assumed you have studied the entire S&T tree and have completed this book. Therefore you will understand that some of the criteria for determining how much to freeze comes from the next couple of steps. In answering the question, "how much is enough?" then, we will consider at least four major elements:

1) At least 25%, perhaps much more
2) The proper corresponding reduction of load on the "Virtual Drum" (see Chapter 9)
3) Enough to optimally assign resources to tasks and projects (see Chapter 8)
4) Enough to assign sufficient resources to full kitting activity (see Chapter 11)

Considering the fact the Virtual Drum sets the pace for the organization, it is most important that it, above all else, is not overloaded. Read and understand Chapter 9 and properly select the Virtual Drum.

Then assess the current load and make sure that when you execute the freeze, the load reduction on the Virtual Drum is *equal to or more* than the general reduction percentage you have chosen. Make any adjustments as necessary.

After determining the number of people necessary for elements 3 and 4, you should create a plan for optimal resource assignment on open projects and full kitting. Remember, these people will be available because of stopping work on the frozen projects. If you do not have enough people for optimal resource assignment and full kitting, *you haven't frozen enough*. Now you should have sufficient information to help you finalize the percentage of load you should reduce.

NOTE: It is better to freeze too much than too little. If you freeze too much and starvation occurs, there is an easy solution—defrost one or more projects. If you freeze too little, you will have to remove more projects from work, assign additional resources to open projects and to full kitting tasks. And you might have to talk to a customer or two a second time.

In summary, the first substantial actions are:

1) Measure the severity of bad multi-tasking.
2) A senior manager in charge of all projects, after consulting with his subordinates, prioritizes all in-work and future projects. Only that manager, or his superiors, if any, have the authority to re-prioritize.
3) Consider the current and future loads on the "Virtual Drum."
4) Using the criteria described above, the senior manager confirms the amount of load to be reduced from work. A line is drawn on the list at the new load level, and projects below the line are suspended from work.
5) By-name lists for optimal assignment of resources to remaining open projects, and for full kitting, are readied for use.
6) Actions are taken to remove frozen projects from work and to ensure they cannot be worked upon until officially "defrosted."

And that's the end of the first day of implementation. Now we will turn to the Three Questions that will energize each step, and help us reach our goal of meeting ALL of our promises, from now on into the future. We will not proceed to the next step until we have adequate answers to these questions. I have included criteria for answering the questions here, but your organization should study them carefully to determine whether they are sufficient, or if you should add additional criteria to fit your environment.

Was the step correctly implemented?

At first glance, this might seem like an unnecessary question. The step was reported as complete. We froze projects. Isn't that good enough? Actually it isn't, unless we have defined the word "correctly". We need to make sure that each of the five actions outlined above were verifiably accomplished as written. Management should not just take people's word for it—they should ask to be shown and make sure they are satisfied before proceeding.

In particular, the fifth action, actually removing projects from work and making sure they cannot be worked upon, is of utmost importance.[20] Physical removal of frozen projects from the work area is ideal, although not always possible. If the project must remain in the work area, it should be visually sequestered and clearly marked that no work is to be performed on it.

[20] Please refer to chapter 12 for the single exception to this rule: completing full kitting for frozen projects.

Figure 11. A properly marked and "sequestered in place" project

In addition, operations routing/testing, etc. paperwork should be removed from the proximity of the project and locked in a secure place. If people have project paperwork, parts, etc. in their desks or lockers, these should be removed as well. If reporting of progress, charges, or time is done on a computer system, the project should be "electronically sequestered" so people cannot access it—it should not be available for such reporting.

NOTE: Critical Chain is applicable across the Projects organization. In some production environments where engineering work is included, and where that engineering work is a significant part of total length of the project, projects in engineering and projects in production can be different, sometimes completely so. Therefore some steps in the S&T tree must be applied in both places separately. If you are in this type environment, it means one prioritizing and freezing list for production and another for engineering.

The "correct" way of implementing this step includes making absolutely sure that no one can access or work on frozen projects. Top management and the implementation leaders should ask to be shown evidence that this is indeed the case. In the words of a 20th century United States president, "Trust, but verify."

Were the expected effects realized?

Here is a great illustration of the "immediate effect" example shown in Chapter 4. Almost without exception, every time I ask a group of people what the expected effects of prioritizing and freezing are, I get answers like "smoother flow," or "more projects completed." Certainly the freeze step contributes to these desired effects, but it is insufficient in itself to cause them. Other actions, such as those following this step, are necessary. So usually I ask the question over and over again until someone allows his or herself to think clearly, and focuses in on the immediate effects. It is important we identify only the immediate effects—otherwise we might miss key additional necessary steps.

Therefore let's consider the true immediate effects of prioritizing and freezing:

1) All employees are aware of which projects are open, and which are frozen.

2) A significant number of these employees find themselves with nothing to do.

It's as simple as that. We want to validate that the step has indeed made people aware of projects they should not work on, and also validate that we now have available resources—at least as many as are on our optimal resource assignment and full kitting personnel lists, often more. If either of these are not the case, we have a problem, and should stop and examine the situation to understand why we have not realized the expected effects. It may be that we have a "mystery" and need to go back and challenge our assumptions.

And although I have included some nominal expected effects at the end of each step in this book, it is important that you as an organization discuss this question for each step so you are clear on the immediate effects (if any) that pertain uniquely to you.

What mechanism has been put in place to ensure compliance with the step over time?

The purpose of the prioritizing and freezing step is to reduce the overload on the system to be more in line with better flow and throughput. But once we have accomplished this, we do not want to let the system ever return to an overloaded condition. We need one or more mechanisms to help keep us from deteriorating from our present low-load state.

The mechanisms suggested here, and the ones you may come up with yourselves, usually involve measurements in one way or another. The only way to know if our flow and throughput are deteriorating is to continually and periodically measure them.

You were already asked to measure the frequency of task switching prior to implementation. Now, once the step has been correctly implemented and the effects have been validated, it is time to measure it again and compare the results. The frequency of task switching should be drastically reduced. Assuming this is the case, you certainly don't want it to rise to damaging levels again. Over time then, a good way to help sustain flow and throughput is to make sure we keep bad multi-tasking under control. On a continual and regular basis (I suggest no less frequently than monthly) we should measure the severity of multi-tasking in each of our departments and watch for negative trends.

In the initial implementation the suggestion is to do an interview, and that will work in the future as well. At a minimum therefore, conduct monthly task-switching interviews and average the results. This time do it by department, so we can see if one department is multi-tasking more than they should.

But we don't necessarily have to do an interview. There are other creative ways to collect this information that do not involve two people (and interviewer and interviewee). For example, you could create a form that employees would check each time they were asked to switch tasks. Every month the forms would be collected, analyzed, and checked for negative trends.

Or it may be possible to build a measurement into a computer system—whenever an employee is asked to switch tasks, he or she enters it into a database. The computer, then can crunch the numbers and identify, by department, the severity of task- switching.[21] Such a system could also produce trend lines and compile information more frequently than once a month, if desired. Just be wary of complicating things with technology.

Another option is to measure deviation from a guiding rule. Let's say your projects are engineering projects, and your product is engineering drawings. Perhaps today each engineer has 20, 30, even 50 projects piled high in his or her cubicle. As a result of the freeze step, a guiding rule is established where the maximum number of projects an engineer may have available is 3. So another mechanism for keeping the organization stable would be to ask how many times there were deviations from the rule. How many times were more than 3 projects available to an engineer, and for how long were they available?

Hopefully you are beginning to see the power of The Three Questions, and how they assist you in executing a successful implementation. But for now those immediate expected effects mean we will have people standing around tomorrow morning with nothing to do. We should have our lists in place and get ready for the next step—Accelerate Project Completion.

[21] If some departments are showing a negative trend toward bad multi-tasking while others are not, it is possible to focus only on selected departments for reduction. Please see Chapter 29 for more information.

Chapter 8 – 5.111.2 Accelerate Project Completion

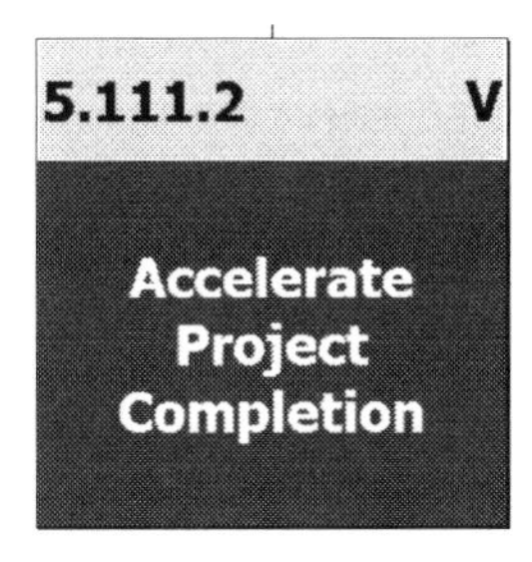

Day two begins. At yesterday's close, projects had been sequestered based on their priority and the desired level of load on the system. As the morning of the second day of the formal implementation kicks off, some projects remain open, and some people go back to the projects they had been working on. Other anxious people are wondering what they will be doing. They have heard about the "lists", and maybe even have seen them. They understand the S&T tree, so they know they will be gainfully employed. No matter how much you have communicated and how well you have tried to prepare for this, there will be a certain degree of apprehension if not downright anxiety on both the part of the leadership as well as project leaders and staff. But with early results forthcoming, anxiety will soon give way to enthusiasm.

Necessary Assumptions

There are an optimal number of resources per task and per project. In most multi-project environments the eagerness to start all projects as fast as they are won causes spreading resources too thin between projects. This practice causes the lead time of all projects to increase and promotes bad multi-tasking.

If flow is our number one consideration and our goal is to become a highly reliable projects organization, we need to take a close look at the proper manning of our still-open projects. Overloading of resources is a clear obstacle to increased flow, but simply reducing the load doesn't give us all the information we need concerning the best way to utilize our people. Therefore, we need a strategy.

Strategy

There is good (sufficient for accelerating project completion) assignment of resources to projects.

The objective is a simple one: complete projects faster. The new strategy is that from now on, rather than assigning minimal resources to projects, we will assign the proper number of resources required to move the project forward with the maximum possible speed.

Parallel Assumptions

Manning of projects according to their optimal number of resources (rather than trying to squeeze in more projects) leads to an overall increase in the rate at which the Company finishes projects while decreasing the projects' lead-times (in some environments by up to 25%). The freeze causes many people not to have an active assignment (and people standing idle spreads demoralization).

Second step, second paradigm shift. The first paradigm shift involved freezing projects already in work and maintaining a lower level of load on the system—one that is better in line with flow and throughput. The second paradigm shift involves how we assign resources to tasks and projects. As hard as the freezing paradigm shift is, for many people this is the hardest one of all. Although decidedly for the better, optimal assignment of resources is excruciatingly hard for many companies. In many Critical Chain implementations that are not fully producing results, you will often find that this step was never properly implemented or was allowed to deteriorate over time.

The main reason this step is so hard is that it really reverses conceptions and behaviors that we have been accustomed to our entire business lives—the concept that once we win the project (or the project has been formally approved, for internal projects), we must get started on it immediately. The pressure is enormous. In addition to the mistaken idea that "the sooner we start it the sooner we'll finish it," there is a feeling that if we don't get started immediately, we may lose the business—we have to be able to demonstrate progress to the customer (for an effective alternative, see Chapter 14). The inevitable result is we tend to assign *minimum* resources to projects, and overload our system. This creates another misconception: we need more resources. It also forces us into bad multi-tasking (task switching), further delaying projects (as all the elephants head for the door at once).

The word *optimal* is derived from the word *optimum*, which is defined as "the greatest degree or best result obtained or obtainable under specific conditions." Using this definition, the necessary assumption above concludes that there is an optimal number of resources that can be assigned to both tasks and projects. This optimal number will produce the best results possible—in this case, the fastest completion of projects possible.

Optimal assignment of resources implies that there can be either too few or too many resources assigned—sub-optimal and super-optimal, respectively. It is clear how too few resources can slow a task or project down, but it is also true that too many resources can do the same—people can get into each other's way and cause delays.

An example I like to use is the painting of a door. A door, hanging on its hinges, has three surfaces—front, back, and edges. It is easy to think of a person painting a door. The painter would start on one side of the door, then perhaps do the other side, then the edges. In this way, assuming each surface takes 5 minutes, it may take him 15 minutes to paint the door. But what if we added a second painter? One painter could paint the front side at the same time as another painted the reverse side. Then one of them could paint the edges. In this way it would take only 10 minutes to paint the door. But what if we added a third painter? With the door open, we could have one painter on each side, and one painting the edges. Now we are able to paint the door in 5 minutes.

Can we add a fourth painter? Maybe, maybe not. At this point, I wouldn't blame you for thinking that the painters might start getting in each other's way, delaying one or more of the other painters, pushing the total time back up above 5 minutes. This is the thinking that pervades most attempts at project acceleration. If you applied this approach consistently throughout the project for all those cases where it makes a difference to the length of the project and you stopped here, you are doing better than most people. However, if you are of the mindset that drives you to go even further in your quest to shorten the duration of your projects, there is another step that should be taken.

Even people who get the basic idea of adding optimum resources sometimes fail at this point to recognize that we still have not fully exploited the potential to accelerate the speed at which we could complete this simple task. Since most tasks today are designed with the assumption that the maximum number of resources that will be available is *one*, very little thought has been given to how the work could be re-arranged to facilitate a productive increase in resources beyond the obvious three in this example.

Why don't you give it a try? Let's pretend that you have done your best so far but the project does not meet the business requirements. Or, pretend you are on a game show and your team has been challenged to cut the duration of your task (paint door) to under 5 minutes, safely and with high quality. Take some time and think about this challenge, and write down all the ideas that come to mind. By the way, this same situation could present itself during execution where the need is not to reduce the initial critical chain but to recover buffer.

So, what did you come up with? How about two people working on each of the top halves of the door while two others work on the bottom halves? Yes, I too can see some potential concerns with this idea. But still, what if it really, really mattered that this task takes no more than say, 3 minutes? Do you think it would be possible to find effective ways to address those concerns? If we could, we could cut the flow by another (roughly) 50% by adding two more people for a total of five. My claim is that it is more than just possible, it is very easy to do! As long as the preparation steps introduced are not on the Critical Chain itself, any time added is acceptable. Sure, we should be reasonable and we should keep an eye on costs, quality as well as safety. However, when flow is truly the number one consideration, we shouldn't be too quick to accept the "obvious" limitations to flow that we encounter. Rather, we should always be asking ourselves, (that is, the key members and supporting cast of the project team), "is this really the best we can do?"

Therefore the rule when assigning resources is to continue to add resources until you can't go any faster. Please note that each additional resource *does not have to increase speed at the same rate,* (e.g. each resource does not have to double the speed) as long as the speed is increased, even incrementally, to its maximum potential while protecting or enhancing safety and quality.

At this point we should address a couple of common objections to this concept. They are in reality based on invalid or *insufficiently-informed* assumptions, but at first they seem quite reasonable. The first is that optimal assignment of resources will add cost to the project. To illustrate why this is not necessarily true, let's consider the following exercise.

A particular task on the Critical Chain of a project (or critical path for those not doing CCPM) takes 40 days when done by one person. Every day that can be saved on this task reduces the overall length of the project, but would it also increase the cost? Which of the options would you choose?

Careful analysis determines that this same 40-day task:

Takes 18 days when performed by two people (36 man-days)
Takes 16 days when performed by three people (48 man-days)
Takes 14 days when performed by four people (56 man-days)
Direct costs are $360 per day per worker (or, per man-day)

What is your answer? Did you choose the four person option? If not, why not? Let's take another look of the costs of the options:

Days	People	M-Days	Daily	Cost
40	1	40	$360	$14,400
18	2	36	$360	$12,960
16	3	48	$360	$17,280
14	4	56	$360	$20,160

Figure 12. The typical "cost-world" view of optimal resource assignment

Maybe your answer was two people. This is understandable because it appears that this option has the lowest cost. It appears that by using two people, we could save $1440 from the one-person option. That looks good! Many organizations wouldn't even take the time to do this calculation, and if they are accustomed to assigning minimal resources they will likely only assign one person, and unknowingly incur the extra cost. On the other hand, if we were to utilize three people we would apparently ring up $2,880 in additional cost from the one person option, and $5,760 additional cost for the four person option. It seems like a no-brainer – that two people is the proper choice. But is it?

It depends. We don't have all the information we need to make the best decision. We have not answered one very important question: *What is a day worth?*

Remember, we said this task is on the Critical Chain (or critical path). Therefore every day we can save shortens the duration of the entire project, and in such a case *time is money*. Let's assume we have agreed with the customer to deliver 200 working days from start, and for this project we will be paid $5 million. Material cost for this project is $2 million, so the throughput[22] for the project is $3 million. Therefore, the value of one day for revenue is $5,000,000 - $2,000,000 = $3,000,000 ÷ 200 = $15,000. By using two people, we save 22 days in project duration. That's 22 days where we can use our resources on another throughput-generating project. So do we just save $14,400 in labor cost?

If a 200 day project is generating $15,000 per day in throughput, and we can shorten the project by 22 days, our new 178-day project generates $16,854 ($3,000,000 ÷ 178) per day. If we can fill those 22 days with a similar throughput project, we could generate an additional $372,227 ($370,787 + $1440) for the company in throughput! Now let's look at adding a third person. By doing this we save two additional days of project duration, down to 176 days. This makes a day worth $17,045 ($3,000,000 ÷ 176) per day. Filling the difference of 24 days with a similar throughput project could generate $406,210 ($409,090 - $2,880 additional

[22] In TOC, the definition of throughput is Selling Price minus Truly Variable Costs (TVC), such as material and per-unit commissions. Labor is generally not considered a Totally Variable Cost. It is variable over time, but not from unit to unit. From our throughput all Operating Expenses are paid. Total throughput, therefore, is a global measure.

labor cost) for the company, or almost $34,000 additional from using only two people.

Finally, let's look at utilizing four people to execute this task. This shaves another two days from the project, and makes the value of a day now $17,241 ($3,000,000 ÷ 174). And filling the newly-available 26 days with a similar throughput project would yield an astonishing $442,516 ($448,276 - $5,760 additional labor cost). This is more than $70,000 more than the two person option. Kind of makes the additional $5,760 seem trivial, does it not? Do you still think two people is the correct answer?

Of course your next project may not produce exactly the same throughput as the first. It may produce more or less additional throughput per day, but the calculations can easily be done, and it's almost always worth it.[23] You just have to think *clearly.* At the very least, when determining the optimal number, do the analysis!

Here is a graphical view of the above scenarios:

People	Days	Days Saved	T-Put/Day	Subtotal	Delta Labor	Opportunity
1	40	0	$15,000	$0	$0	$0
2	18	22	$16,854	$370,787	($1,440)	$372,227
3	16	24	$17,045	$409,090	$2,880	$406,210
4	14	26	$17,241	$448,276	$5,760	$442,516

Figure 13. The true value of optimal resource assignment

The second common objection begins by looking like more of an obstacle than an objection. It sounds something like this: "We can't assign optimal resources, because we don't have enough sufficiently skilled people[24]." Digging deeper, we discover the objection, and the specific fear

[23] Even if every day saved cannot be filled completely, you can still experience great value. The hidden capacity you are revealing can be used for cross-training, certifications, preventative maintenance, etc. It's a win either way!

[24] First, I need to ask you one more time: are you sure you have frozen enough? However, sometimes skilled resources are extremely hard to find, and certain skills may

that motivates it: "If we assign less than highly-skilled people to a task, they will make mistakes, cause rework, and extend rather than shorten the project."

It is true that we may experience unpredictable negative effects from assigning less-than-perfectly skilled people to tasks. But both experience and computer simulation are indicating the negatives are not as severe as we may expect, and are quickly offset by the incremental improvements contributed by the "new" people . Keeping in mind the value of a day, any help at all that shortens the project is highly positive. And the benefits to the company are priceless as we create a better cross-trained workforce continually growing in greater competency and skill.

Except in cases where certifications are required before work can be done, adding less-skilled resources to tasks is almost always a good idea. When someone justifies not assigning resources based on nebulous and unquantifiable fear, they are also condemning the organization to the perpetuation of the problem, since the resultant growth in skill of additional resources is extremely hard to come by when, as we know from the current situation there is "little or no time for training."

The Parallel Assumption suggests that the lead time of projects using optimal assignment of resources can be reduced up to 25%. Experience is now showing the reduction can be even greater. And when you add in other steps from the S&T tree, such as full kitting (having the paint, brushes, and masking tape ready in advance, for example), project lead times can often be reduced by 50% or even more. Now, with the people newly available to us as a result of the freeze step, we can optimally assign resources.

Tactics

The optimal number of the various types of resources needed for each open project is determined. The freed resources are used to prudently strengthen the open projects. Proper manning decisions are also done for the frozen and to be released projects.

---

be legitimately "out of balance" with other resources. However, the concept suggested here helps to mitigate this imbalance.

In our preparations for implementation we have prepared a by-name and skill list of available resources. We can now assign these people to our remaining open projects. We will assign people using our best early consensus on optimal numbers. We can make adjustments later if we find that some tasks are either still sub-optimally or now super-optimally assigned. In fact, we are quite likely to find out now that we *legitimately* have too many or too few of certain resources. This is almost always the case, which is understandable because of two related reasons: First, rampant multi-tasking makes it hard to get a clear picture of our true needs, and second, very few organizations take the time and effort required to understand the true nature of their resource needs. In those cases where they do, these needs change frequently, due to both volume and project mix, making it difficult to maintain an accurate picture over time.

Count on getting new insight into what your idealized resource picture should be. Count on this picture changing over time as the work ebbs and flows and changes. Your HR organization should be enlisted to help address the longer term implications of the imbalances you will encounter. In the short term, you to decide what to do when faced with shortages some areas and overages in other areas. This is another situation where many organizations falter in their Critical Chain journey. If not recognized and addressed effectively and decisively, the expected acceleration will not happen.

Two anti-flow forces are at work here. The first is the pressure to keep the resources of which we have too many *busy all the time*. The way this is usually done is to violate the edict to keep work in progress at the newer, low level and to release/start more projects. But these projects *also* need the time and attention of the resources for which we do not have enough. This unleashes the second force, the pressure for key resources (usually those we have too little of already) to work on these new projects "just enough" so that there will be something for the other resources to do. The multi-tasking is now officially sanctioned by management, because in their eyes, "surely, this is for a good cause, right?"

But the false idea that we can accomplish more if everyone is always busy is in reality local optimization destructive to the overall system's productivity.[25] It should be clear to the reader that asking already scarce resources to work on *more* things cannot possibly lead to project acceleration. The conflict arises because of the lack of awareness or the unwillingness to accept that it is practically impossible in a project environment to have a precise match between workload and resources of various skill and experience levels. Unfortunately, this bit of reality becomes painfully clear, sometimes too painfully so, once WIP has been reduced. In preparing for the start of the implementation, it is important to ensure that the key decision makers are adequately prepared to address the resulting pressure to launch more work back into the system.

NOTE: Flow is the number one consideration. We never use the Theory of Constraints to down-size or "right-size" organizations. Rather, we believe in increasing sales and throughput through increased performance, reputation and meeting clients' significant needs in a way no significant competitor can. Cutting resources to a bare minimum means we are subject to delays based on unforeseen events. To grow and flourish, Protective Capacity—some capacity above optimal resource assignment and full kitting—is essential. Do it right and you'll grow so much you'll need more employees—not fewer.

Remember, you should still have enough people left after optimal assignment of resources to staff full kitting, plus a few more for protective capacity. If you don't, you *still haven't frozen enough*. Now we will again turn to The Three Questions to see if we are ready to proceed to the next step.

Was the step correctly implemented?

With the last set of Three Questions we should already have validated that the by-name and skill list of personnel were actually available. Now we need evidence that we have indeed assigned those people to open projects where they are able to speed up completion of

[25] It is even worse (much worse) if a company measures people by the degree of their utilization or "efficiency." This institutionalizes the forces which cause local optimization and actually penalizes people for doing the right thing. Therefore they are always incentivized to do the *wrong* thing. And the truism often used by Dr. Goldratt is supremely useful here: "Tell me how you measure me, and I'll tell you how I'll behave."

tasks.[26] Once we are satisfied, we will turn to the frozen projects, and make sure to plan optimal resource assignment for the remaining uncompleted tasks when we return these projects to work. Finally, we will assess upcoming projects and make sure we plan for optimal resource assignment on those as well.

The step is correctly implemented when we are shown that all three categories have been planned for optimal resource assignment.

Were the expected results realized?

Of course the primary thing we are looking for is evidence that within a few days tasks indeed are finishing noticeably faster than they were before the optimal resource assignment. It won't be true for every task because of normal variation (a machine may break down, someone may call in sick, etc.) But it should be apparent that things are generally moving faster. We may also have one or two projects nearing completion faster than we would have expected otherwise.

In addition, some nice side benefits should be improved communication in the work area and a general sense of excitement that something is really happening – finally! If we don't see these things we need to quickly evaluate why, and we may need to revisit our assumptions. Remember that **"to ensure an outstanding start of a major initiative it is vital that the first substantial actions will result in immediate substantial benefits."**

What mechanism has been put in place to ensure compliance with the step over time?

Documentation of optimal resource assignment by task is essential for quick resource assignment when staffing identical tasks in the future. Over time you may adjust these a bit, but let's get started right away. In a

---

[26] Some people prefer to work alone, and may object to having someone assigned with them, claiming adding resources will not make their task faster. But if the consensus is that it will, these people should be instructed to give it a try. Unfortunately, they may have to be monitored to make sure they don't behave in a way that makes their contention a self-fulfilling prophecy.

future step we will build templates for recurring or very similar projects, so making these determinations now will be valuable when that time comes as well. This will help us build a mechanism for ensuring we always have optimal resource assignment.

One suggestion is to have someone periodically check that the standards we are building are actually being executed from project to project. This is also a good time to determine by consensus if adjustments are needed. Deviations from optimal manning of resources should be noted and watched for improvement or deterioration.

Chapter 9 – 5.111.3 Defrost Mechanism

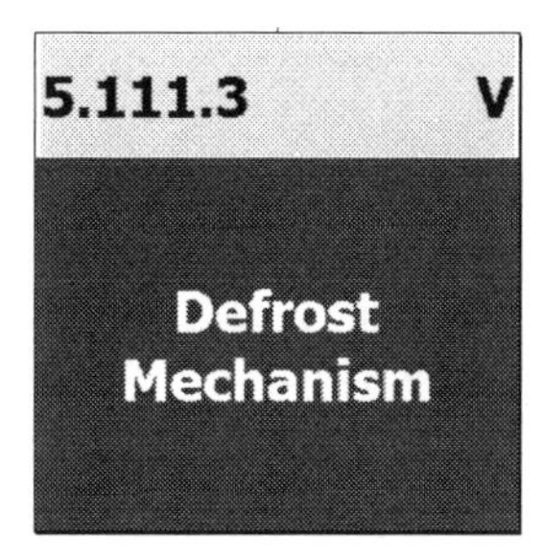

As a direct result of placing the elephants in procession, with a lower Work-In-Progress level and optimal assignment of resources, projects will begin to complete at a much-improved rate. Rather than our projects being abandoned – encased in glacial ice in Antarctica forever – we quickly see that "freeze" really just means "a temporary pause." We have frozen projects in order to go faster, and even the last elephant will leave the room earlier than it would have otherwise. Now let's look at why we need a strategy for defrosting projects.

Necessary Assumptions

Defrosting projects too early will, again, flood the system with work. Defrosting projects too late will lead to starvation of work and unnecessarily extend projects' lead times.

Understanding that a balance needs to be maintained—we do not want to overload the system again, while starvation is also something we want to avoid—we need a strategy for consistently defrosting frozen projects and re-inserting them into Work-In-Progress.

Strategy

Frozen projects are defrosted at a pace that maintains the reduced load.

Although the act of freezing is a one-time event[27], maintaining the reduced load on the system is of paramount long-term importance. The only situations where you should ever add additional load to the system are:

1) You have frozen too much and are experiencing "starvation" (very rare).
2) You are improving so much you can adjust the "Virtual Drum" upwards (see below, also Chapters 18 and 24).
3) You have so much new business you are adding resources (while being careful to maintain Protective Capacity across the system).

For now, we will only consider maintaining the lower level of load on the system.

Parallel Assumptions

The level of the reduced load is approximately maintained when defrosting projects is in-sync with projects being completed. Defrosting projects in-sync with the link that determines the pace of projects' completion also provides focusing on which actions/initiatives help and which jeopardize the flow. In multi-project environments the factor that determines the pace of project completions is not the most loaded department but the synchronization between the various "legs" of the projects. Integration is the link where, for each project, the various legs are coming together. Having too many projects in integration diffuses the efforts to complete projects according to their priorities since whenever a problem that requires chasing a resource from another department is encountered the tendency is to work on another project.

It used to be believed that the heaviest-loaded department was the thing that determined the pace of project completion in a multi-project system. This was a carry-over from TOC for Production Operations, where we were always concerned about Capacity Constraining Resources, or CCRs. It certainly made sense and using the most heavily loaded resource as the "drum" or pacing link to projects completion, did produce result

[27] Just be sure you have frozen "enough." Having to freeze a second time because you are still overloaded can be a difficult and embarrassing experience.

much of the time. However it did not always fully produce the results we desired, i.e. project environments where on-time delivery performance was 95% or greater. In other words, although we were using the most heavily loaded resource to "drum" the system, often we realized that *something else* was exhibiting even more control over the pace.

With years of experience and analysis, it was finally determined that the real pacing element in project environments is synchronization between "legs" of the project—the points where sections (chains) of work come together to join other chains—especially the Critical Chain. In order to avoid delays, these sections of work have to be synchronized—the right things must be in the right places at the right time—not an easy thing to accomplish. What's more, this element is not entirely physical—it is a *phase* in the project made up of various tasks and resources. This also means it cannot be represented in a PERT chart or project network as a single task or resource.

Given that the phase involves sections of work coming together and entering the Critical Chain, we refer to it as Integration.[28] The notional representation of a simple project in Figure 14 highlights the Integration phase of the project.[29]

Since it is not entirely physical, we refer to this pacing element of the system as a "virtual." Linking the pace of the "virtual" drum to defrosting of projects will approximately maintain the new lower level of load on the system.

---

[28] Some organizations already use the term "Integration" for other things, such as final assembly where large products are produced as projects. It is important when explaining this to others that they understand the distinction in meaning from a TOC/CCPM perspective.

[29] Since the "legs coming together" phase may vary from project to project, the integration phase is identified *by project*. But very often the integration phases of projects will be identical or nearly identical.

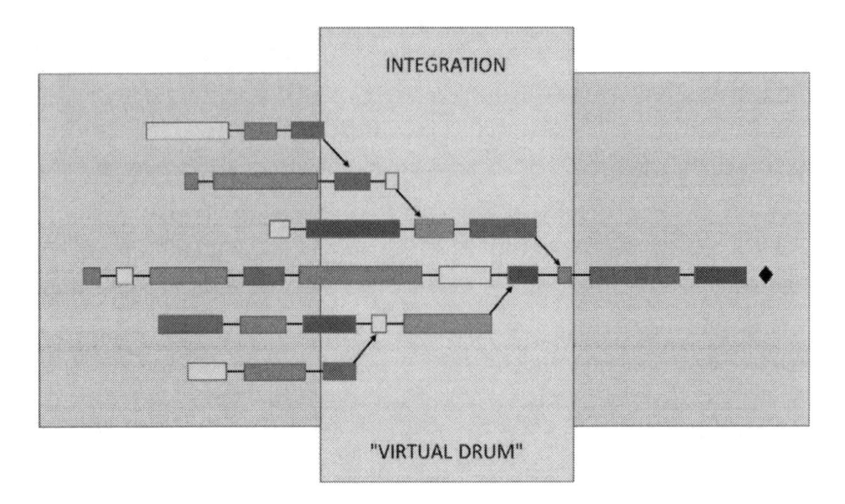

Figure 14. The full Integration Phase represented as the Virtual Drum

Tactics

The company chooses integration (or part of it) as the VIRTUAL DRUM: The number of projects allowed in that section is restricted to be, at most, 75% of the current number. When a project completes this integration a frozen project is defrosted. The sequence of defrosting projects is according to the agreed projects prioritization.

The Virtual Drum is valuable in both planning and execution. By utilizing the Drum concept in our planning, we are attempting to smooth the flow of the system and eliminate (as much as possible) resource contention across projects. Therefore we select the phase that paces the system (the area of synchronization—Integration) as the Virtual Drum and limit the number of projects we plan to be worked upon concurrently within that phase.

As a "first cut" at using Integration to set the pace of the system we ensure that the reduction in projects within the Integration Phase is equal to or greater than our chosen general reduction in the Prioritizing and Freezing step, which is always at least 25%. In execution, we monitor the Integration Phase for load level and for the completion of projects. As soon as a project completes the Virtual Drum phase, another project is defrosted, or released from the freeze state. This keeps the load on the system relatively stable at the new, lower level until the Critical Chain steps can be taken.

Notice that the tactic says "the company chooses integration (or part of it) as the Virtual Drum." Since it is always a good idea to try to isolate the actual pacing phase so we don't try to make adjustments "within the noise," we should understand the criteria for selecting "part of" the integration phase as opposed to all of it. In some environments, the integration phase can be quite long – weeks to months long, for example. In any case, you should consider the possibility of selecting only a part of integration as the Virtual Drum. But you cannot just select randomly. Remember, we are using the Virtual Drum as a pacing mechanism. Therefore it must match the actual pace of the system. The idea is that both upstream and downstream of integration, things generally flow faster. But this can also be true even *within* integration.

The Virtual Drum is like a deep pool of slow water in an otherwise swift river. In some environments, we can identify a section of integration that is generally slower than other sections—a place where problems more often occur—where the engineers tend to be called in—where management must focus their attention in order to keep the project moving. If such a clear dichotomy exists within the integration phase, you may divide it and use only the "slower" section as the Virtual Drum.

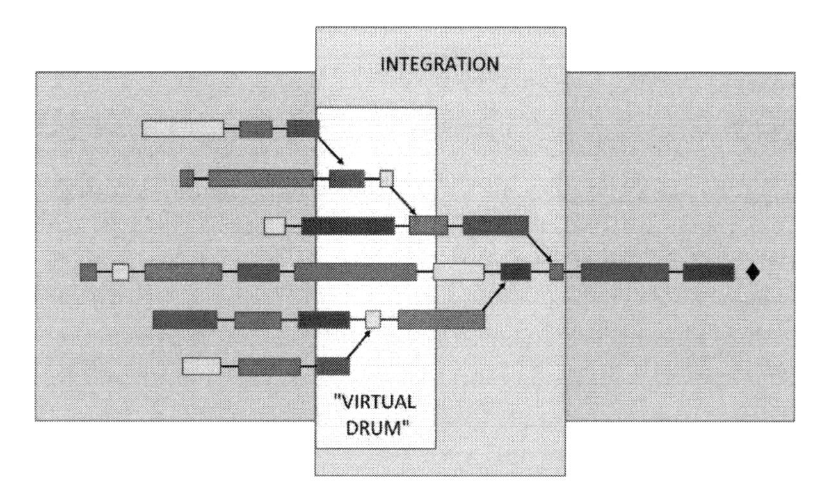

Figure 15. Part of the Integration Phase selected as the Virtual Drum

Careful consideration should be given to how many projects should be allowed concurrently in this phase—it will be a smaller number than would be allowed if the entire integration phase was used as the Virtual Drum. Also, the word "complete" should be clearly defined. What exactly does it mean when you say a project has "completed" the Virtual Drum phase? Usually it means that all the work planned for that phase has been successfully completed, but you can define it further. A frozen project has not truly completed the phase until all approvals and inspections have been successfully concluded. It is very important that we do not defer or "travel" Virtual Drum work to other parts of the project. Nor do we want to be doing rework later on because we declared "completion" prematurely.

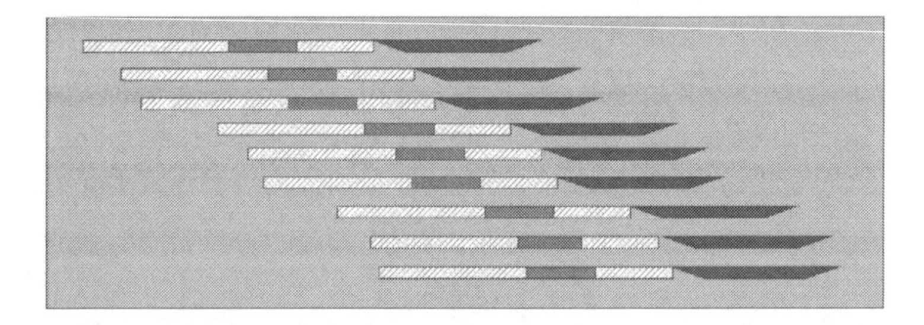

Figure 16. Notional view of a stagger of several similar projects where orange represents the Virtual Drum and green represents the project buffer. In this example, only 3 projects are allowed concurrently in the Virtual Drum phase.

Finally, as a project completes this phase, a frozen project (with full kit only)[30] can be defrosted. In order to "defrost" a project, specific administrative processes must be in place to indicate to all affected parties that it is now acceptable again to work on this project. For example, if the time reporting system was used to prevent team members from reporting time against this project while the project was frozen, this process would now be reversed and the reversal announced publicly. When a project is "defrosted" steps should be taken to confirm that given potential changes in personnel, there are no important task or coordination requirements that have been overlooked. It is a good idea to hold a mini-kick-off meeting

[30] See Chapters 11-13 for a complete discussion of the full kit. In the most effective implementations, full kit includes available resources for optimal assignment.

to familiarize everyone (especially new team members) with the project and to confirm what was already completed and what is still left to be done.

Remember that projects should be defrosted only in priority order. Now let's consider the Three Questions in relation to this step.

Was the step correctly implemented?

Among the things to verify:

That integration was properly defined for each project

That if "part of" the integration phase is identified as the Virtual Drum, the selected "part" is actually the most problematic area of the integration phase

That "completed" has been properly defined

That the organization receives a clear signal when a project completes the Virtual Drum phase

Were the expected effects realized?

The expected effect of the defrost mechanism is quite simple: frozen projects are immediately defrosted and returned to work when a project completes the Virtual Drum phase, and not before. Eventually, all previously frozen projects should return to being active, without any aspect of a project falling through the cracks due to poor process coordination. We expect the overall load on the system to be reasonably stabilized[31] going forward.

What mechanism has been put in place to ensure compliance with the step over time?

A mechanism is needed to sustain the defrost mechanism until all projects are defrosted and the Critical Chain steps modify the release of new projects into the system. One suggestion is to monitor the signal that a

---

[31] Since projects vary in size, load on the system can be somewhat variable. But generally the load should be stabilized, and should not be showing wild oscillation.

project has completed the Virtual Drum and measure not only the fact that another project has been defrosted, but the length in time from the signal to the time the defrosted project is actually ready for work. Watch for trends and find ways to reduce the timeframe. Another suggested mechanism is to measure the pace at which projects are completing, once again watching for trends (remember we watch for trends because a single data point or two could be an aberration). The expected effect is an ever-increasing pace due to the improvement in your performance. If this effect does not occur or the pace slows down, the system should be examined for the reasons.

Chapter 10 – 5.111.4 Releasing of New Projects

NOTE: This is a transitional step. The defrosting process takes some time – usually several weeks – and although it is hoped that during this period you have been able to complete the construction of a full kitting system (4.11.2) and implemented Critical Chain planning and scheduling (4.11.3), this is not always possible. Therefore you may release new projects for a while in the same manner in which you defrosted projects, i.e. when a project completes the Virtual Drum phase you release a new project into work. But this step is not truly complete until both 4.11.2 and 4.11.3 have been completed and you are releasing projects by "leg" via Critical Chain software. Just remember to work diligently on completing these steps, because flow is the number one consideration!

Sooner than you might think, it will be time to begin releasing new projects into the system. During the time that you have been defrosting frozen projects, there has been a lot of activity taking place. You should have a Full Kit Manager and a robust preparations system in place (see Chapters 11-13) and more. Now it's time to release new projects under the new paradigms, and for that we will consider the strategy and tactics of doing so.

Necessary Assumptions

For most projects there is vast difference between the lead- times of their various "legs"; there is no one date for release of a project. Release of all legs of the project at one shot increases unnecessarily the load. Note: For frozen projects most "legs" have already been released.

Keeping in mind that flow is the number one consideration, we always need to be vigilant in not overloading our system and ensuring the greatest possible velocity of projects. Therefore as we begin to introduce new projects into our system, we need a strategy that puts flow first.

Strategy

The timing for the release of each "leg" of a new project takes into account the lead- time of the leg.

The first thing we learn here involves yet another paradigm shift. All projects generally have multiple paths, or "legs", that merge together at various points in the project, finally leading to a single path and project completion. In the past, when we started a new project, it is very likely that work began immediately at any point in the project that was ready to go. Sometimes that meant we would start every leg of the project at once. In our new paradigm, that will no longer be the case. We will no longer start all legs as soon as possible, but rather we will start each leg only as required by schedule.[32] This is another safeguard against overload and the resulting task switching. As before, this lower level of load further enables speed in the projects. This also allows us to do full kitting by leg, smoothing the load on our full kit resources.

Releasing a new project, therefore, really means starting just the "longest leg" of the Project, which is almost always the Critical Chain path. With this understanding, it is possible in some situations to release the longest leg of a subsequent project before releasing the shortest leg of the current one.

Parallel Assumptions

For most multi-project environments it is too cumbersome to manually calculate properly the release dates of the various legs of new projects. Most project environments (and most commercially available software) do not consider the fact that the lead time of the various "legs" of a

[32] In some unique environments there are exceptions to this rule. However for this discussion we will stick to the generic case.

project are also a function of the load on the various resources (Critical Path vs. Critical Chain - removing resource contentions). The lead time of a project and the lead time of the various legs of a project are a function of the way safety is included (safety in the task level or in the project level - Project and Feeding Buffers). Most project environments (and most commercially available software) do not use the concept of Project and Feeding Buffers.

At this point, well over 100 pages into this book, the S&T at last introduces the subject of software. This should help dispel any notion that software is the solution – in fact software is never the solution. However, due to the sheer numbers and magnitude of the calculations required to release projects by leg, we could reasonably say that to successfully complete this step, software is necessary. In addition, the parallel assumptions tell us that most commercially available project management software products do not consider resource contention in their calculations and do not use the concept of Project and Feeding Buffers. Therefore specialized software is required for CCPM.

Several different CCPM software products are available. Some are better than others but all provide the minimum set of CCPM features you will need. CCPM software products all identify the Critical Chain and insert necessary buffers. The best available products have "what if" capability to help predict the impact of schedule changes or emergent projects on the system, the *ability to identify the Virtual Drum phase by project* and provide task priorities across projects. You may also consider products that help with full kitting and continuous improvement (POOGI capability), products that are available on the cloud, etc. Their features and capabilities should also be consistent with the requirements of the Projects S&T Tree.

Tactics

When the time arrives to release new projects, steps 4.11.2 and 4.11.3 should be in place. At that stage, a system to release new projects using the CCPM concepts is ready.

What might appear to some as an exception to the step-by-step rule is really not an exception at all. Once freezing has occurred and optimal assignment of resources has begun, they can be thought of as complete (actually they continue on in maintenance mode). Full Kitting (one functional manager) and project network construction (a different functional manager) can immediately commence. Building the system and documentation for Preparations is the subject of 4.11.2. Critical Chain Planning and Buffering is the subject of 4.11.3. By the time all frozen projects have been defrosted and you are ready to release new projects, 4.11.2 and 4.11.3 should be essentially complete. After the final frozen project has been defrosted, as soon as the next project already in work completes the Virtual Drum phase, new projects should begin to be released in priority order.

When considering the action component of the tactic, pay special attention to the second sentence: *"At that stage, a system to release new projects using the CCPM concepts is ready."* What is this system? While frozen projects were defrosting, we have been putting it in place. In addition to the Full Kit steps mentioned at the beginning of this chapter, we have also been completing the Level 5 entities under 4.11.3:

1) Building good project plans/PERTs (Chapter 16)
2) Building Critical Chain Plans (Chapter 17)
3) Staggering Project Portfolio (Chapter 18)

The figure below shows a rough view of the approximate sequencing and timing of the S&T steps to this point. A more detailed depiction can be found in Chapter 30. Please note that installing and stabilizing the software, building processes, training resources by role are pre-requisites to 5113.2/3. At this point we will also begin execution of the day-to-day operations, roles, and responsibilities outlined in the next several chapters.

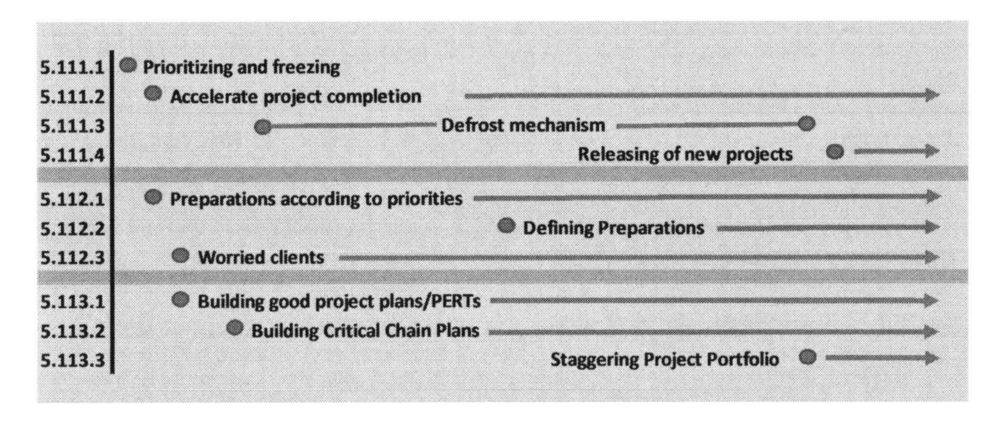

Figure 17. Sequencing and approximate timing of S&T steps to this point

Was the step correctly implemented?

This is a good time to internally double-check everything that has gone before, ensuring that freezing, optimal assignment of resources, full kitting, project network construction, Critical Chain Planning and Virtual Drum staggering are all without significant deficiencies and are operating smoothly in the CCPM software. By this point you should also be seeing major acceleration of projects as compared to the old system, a quicker pace of completions, and an increased percentage of on-time completions. Although you should have been benefitting from the services of an Auditor during the implementation, this is also an excellent time for an official audit of your implementation by the Auditor's unbiased, outside eyes. I highly encourage you to ensure this occurs, to professionally assess everything and make sure you are on the strongest possible path toward success.

Were the expected effects realized?

The expected effects of correctly implementing this step are a working system of Critical Chain planning and buffering which via CCPM software automates much of what you have been doing during the transformation from traditional to CCPM systems. All projects should be consistently full-kitted, optimally staffed and released according to their schedules. (Earlier is OK as long as there is no multitasking,) Resources are able to complete work and rework on other projects before being forced to jump on new projects. Within larger projects, multi-tasking between legs is consistently reduced.

At this time you should also have significant buy-in and enthusiasm from the entire organization for sustaining the system well into the future. If you have not experienced these things, you need to do an analysis to find out why, and make the necessary corrections.

What mechanism has been put in place to ensure compliance with the step over time?

You might be tempted to say that the CCPM software you have activated is the mechanism which ensures you will release new projects properly from now forward. But the software is just an aid to indicate when you should release (the longest leg of) a new project – it does not ensure that you do so. Therefore you should continue (or modify if necessary) the measurements you created for defrosting in the last chapter, and monitor new releases in relation to Virtual Drum completions. How long after completing the Virtual Drum phase did you take to release the next project? What is the pace of your new releases? Are there any negative trends over time? Are you getting so good, you are starting to see "holes" in the pipeline in the future? Remember the warning in the "How to use this book" section about becoming super-productive, and make sure your sales people have created and are actively pursuing a great TOC market offer to keep the pipeline full.

Section Three: Preparations

Flow is the number one consideration

Chapter 11 – 4.11.2 Full Kitting

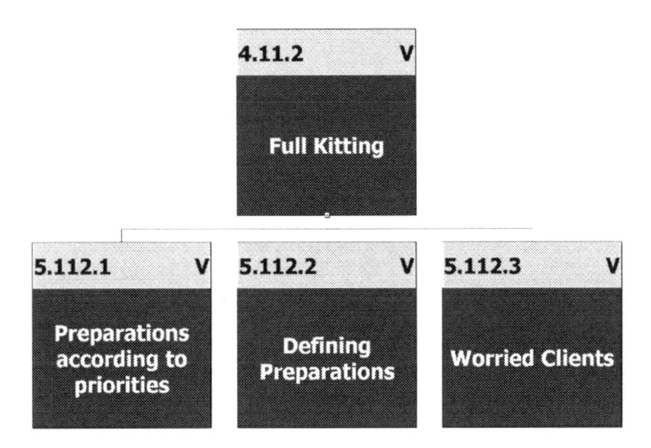

Once we have successfully completed freezing, we can move quickly, as we did with acceleration via optimal assignment of resources, into preparations, or Full Kitting. But planning for both these activities can and should begin earlier, prior to the actual execution of prioritizing and freezing. For acceleration, we can review existing project plans for work not yet completed and determine what the optimum staffing would be for each task. For full kitting, we can appoint a Full Kit Manager (Chapter 12) and determine the personnel requirements for a permanent full kitting operation (Chapter 13).

Freezing and acceleration remove the delays caused by overloading our resources and its resultant task switching. But in addition to the delays caused by bad multi-tasking, there are other delays that are just as devastating to lead time and due date performance – delays caused by the organization having to put projects already in work on hold while they wait for something – a key purchased or subcontracted material, a customer approval, a piece of crucial engineering, or one of many other things. These delays also contribute to pressures to overload our system once again by releasing other work too early, with the well-intended but deleterious objective of resources "always having something to work on", which will eventually send us back into chronic task-switching.

The next big step in improvement comes by making sure we have what we need when we start a new project. It means completing our preparations. Having what we need is the same as saying we have a "Full Kit." In order to eliminate delays resulting from missing items, we need a strategy and one or more corresponding tactics.

Necessary Assumptions

The current pressure often causes projects to be in execution without the needed preparations being completed (detailed specifications, authorizations, etc.).

There are many reasons we are tempted to start projects before all preparations are in place. One major reason is the erroneous unstated assumption that the sooner we start a project, the sooner it will finish. However, we also know that the more projects we are working on concurrently, the longer each project takes to be completed. Furthermore, as we sacrifice preparation activities in order to get going quickly, we are compounding the situation as we most likely will find ourselves blocked due to missing inputs later in the project.

A common argument in defense of the decision to start all projects ASAP is to blame it on the customer's desire to see immediate progress or to trigger progress payments from the customer sooner rather than later. This concept, and a better alternative, are discussed in Chapter 14. While these arguments may have merit, the fact remains that they lead to all projects taking longer to complete.

Additionally, sometimes there are very concrete reasons – real pressures – to start without "full kit." If we are heavily loaded and missing many of our due dates, we may feel we have "no choice" but to get the project started in order to have even the slightest chance of completing on time. This often happens when sales departments commit deliveries without coordinating with production, a major client has a crisis and creates an emergent order, a customer issues a change order but the completion date is not adjusted, or we want to impress a new client prospect in order to gain new business. There are even some cases (such as with government contracts) where the customer *demands* to see progress commencing immediately upon contract signing. With some customers, it almost feels like it's more important to them that their project is started than it is that their project is finished.

At the same time, everyone understands the value of being prepared, and we often hear voices within the organization calling (or begging) for full kit, due to the chaos caused by having to continually stop and start work due to missing inputs. But even if an ad-hoc full-kitting policy is imposed, it quickly breaks down as the pressures start to build once again to deliver on time. There must be a better way.

Strategy

A project is rarely launched before its preparations are complete.

The first thing we notice is that the strategy says projects are "rarely" launched before full kit. It does not say they are *never* launched before full kit. However, let's look at the dictionary definition of "rarely", because the word is used deliberately and it is a crucial word:

Rare – coming or occurring far apart in time; unusual; uncommon.

It is important that we be literal here. Rare means rare. Releasing projects without full kit should not be common, nor should approval be easily obtained. Having full kit before we release projects into work should be common - the default condition of the company – business as usual, so to speak. It is the exceptions that should make us nervous, not the rule.

This is the "what" of the step. For the "how" of the step we need one or more tactics. And in order to select the best tactics, we need to take a look at the Parallel Assumption(s).

Parallel Assumptions

The resources dealing with preparations are caught in a never-ending catch-up cycle. Freezing of projects frees up, for a while, ample capacity of the resources dealing with preparations.

Due to the pre-implementation state (our old "business as usual") - the earlier chaotic situation, we find ourselves always behind in fully preparing for projects. We have shortages and delays everywhere in our current, released projects. We need to *catch up* everywhere before we can even think of maintaining a full kitting program.

Tactics

The company uses the window of reduced load on resources that do the preparations to ensure that "full kit" practice will become the norm.

Before we specifically consider the tactic, a word of clarification: In the last chapter, when we were discussing the release of new projects, we said that projects should be released *by leg* of the project, rather than releasing everything at once. Likewise, full kitting for projects is done by leg, with the longest, or Critical Chain leg, coming first. Since it is the Critical Chain, it is vital that this chain, or leg, should not be delayed for any reason. However, this means that we do not have to wait for full kit on every leg of the project before we can release the project.

On the other hand, it is important that the other legs of the project are also released with full kit, because delays on a feeding chain, if severe enough, can begin to penetrate the project buffer, making a feeding chain the most-penetrating chain. Therefore we must not needlessly let feeding chains be consumed – we need to pay attention them. We need to stay "on top" of purchase orders, engineering, etc. and take action if our start dates are threatened. But in the big picture, full kit on the Critical Chain is of utmost importance. We have a bit more buffer elsewhere.

We now return to the tactic. A result of freezing is the freeing-up of resources. With some of these resources, we have assigned people in an optimal manner to tasks and open projects. If we have frozen correctly – literally "enough," – we have sufficient left-over resources for building and maintaining the full-kitting program. Although maintaining the program will require some permanent resources which should be identified and trained as early as possible, swift construction of the program exploits (or takes advantage of) the window of time/resources created when we initially prioritized and froze projects. It is crucial then that we begin acting immediately in concert with freezing, and work diligently to build and staff the full kit program in order to establish and maintain our new "business as usual," and then we can begin to reassign our extra resources to other areas.

Once the initial backlog of active and frozen projects have been full-kitted, we should update our full kit policies, procedures and checklists and finalize the logistics and resources necessary to consistently perform full kitting on all new projects. Some resources that were initially dedicated to full kitting activities may now be available to join projects in execution or to become members of the network development / templates / CCPM planning team.

Sufficiency Assumptions

An exception to the rule might be misused in order to by-pass the rule.

Since it is obvious we don't have enough information to construct a program to insure all preparations are ready, we would expect a sufficiency assumption here, with more information on how to accomplish it on Level 5. However, in addition to this need we are given a warning – it's all well and good to have a preparations program, but unless we have strictly defined rules for exceptions[33] to starting projects without Full Kit,

---

[33] What constitutes a "reasonable" exception? The final definition should be made by your management team, but let's use the example of a promised purchased part that has yet to arrive as the start date for the Critical Chain approaches. In my mind, having a "will-ship" date from the supplier is definitely not sufficient to justify an exception. If however, the supplier verifies the part has shipped, with manifest and tracking numbers, and the shipping company has quoted a delivery date, it might be grounds for an exception. However, what is the history of this or similar parts in receiving inspection? Are there frequent rejections involved? And if the part is being shipped internationally, what is the chance of it being delayed in customs? All these things and more should be considered when constructing an exception policy.

exceptions could be used to go around the system, and it could become chronic, returning the system to shortages and delays.

In order to ensure that "rarely" is the actual rather than just the ideal, a documented policy/procedure should be maintained, detailing the circumstances for which approval *might* be granted, the required signature approvals (beginning with the Full Kit Manager), and perhaps even a numerical limit on the number of exceptions allowed at any given time or during any given time frame.

For all this information and more, we need another level of strategy and tactics.

Chapter 12 – 5.112.1 Preparations According to Priorities

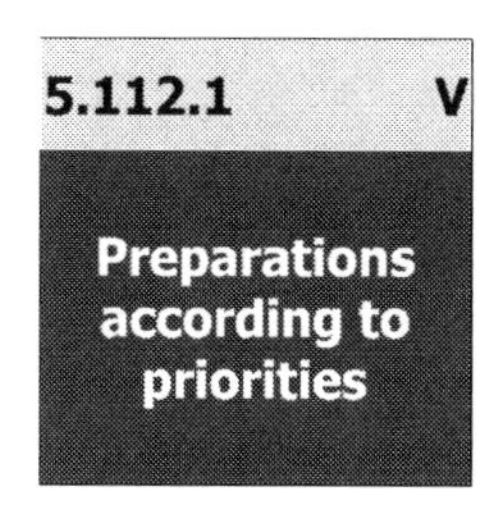

A common question that I hear when introducing what the S&T has to say about preparations is, "Why does 'Preparations according to priorities' come before 'Defining Preparations?' Shouldn't you define and document your preparations before you launch into working on them?" It's a good question, but it is one that has a very practical and carefully reasoned answer:

Flow is the number one consideration.

Once you have frozen projects, it is very likely the projects that remain open still have some shortfalls in preparations. This is understandable because prior to implementation, full kit was not a top priority for starting a project. It is just as likely that your frozen projects have similar shortfalls. And since the projects in both categories have already been in work, it is likely you already know exactly what those shortfalls are. Therefore the best way to accelerate project completion after reducing task switching is to reconcile these shortfalls as quickly as possible.

Then, while doing the defining and documentation, you can continue to reconcile the shortfalls of the soon to-be-released projects. In addition to accelerating this group of projects, we are also contributing to the success of the idea that:

To ensure an outstanding start of a major initiative it is vital that the first substantial actions will result in immediate substantial benefits.

So preparations according to priorities comes first. To do this in the best possible way, we need a strategy.

Necessary Assumptions

In most multi-project environments the importance of complete preparations - "full kit" - is frequently/constantly radiated by top operational managers. The mere fact that delays and even rework caused by missing preparations are so prevalent indicates that usually the drive to "full kit" quickly deteriorates to lip service.

As was indicated in the last chapter, the idea of full kit isn't new or uncommon. However the practice of full kit is as elusive as the unicorn. Everyone agrees it's a good idea, but the pressure applied by management, customers and others can be enormous, and as the assumption says, "The drive to full kit quickly deteriorates to lip service."

Being aware of this situation, what strategy do we need to make full kit a reality? Is it possible to demand full kit and still make our due dates?

Strategy

Resources and project leaders are used to working on projects whose preparations are (almost) fully completed.

The words "used to" mean that people are accustomed to things to the extent that they don't expect anything different, and it is a surprise when something different occurs. For example, when we enter a room and flip on the light switch, we are "used to" the lights coming on. In the rare cases where the light is burnt out or the power is down, we are surprised by the change from the norm. This is the level of full kit performance we are striving for through our preparations strategy.

But there is a qualifier word here, as the strategy states we are "used to working on projects whose preparations are *(almost)* fully completed." "Almost" does not give license to randomly or negligently release projects without full kit, however. There are three controlled factors that might mean the kit can be released without being completely full. These are:

1) The Critical Chain leg indeed has full kit, but one or more of the feeding chains still lacks an item or two.
2) An official exception with signature authority has been obtained.
3) The full kit is compliant with a minimum level of completeness (as described in the tactics section below.)

Now the question becomes, "What are the best tactics for reaching this strategy?" For that, we look to the parallel assumptions.

Parallel Assumptions

A powerful way of turning a good mode of operation into the norm is to ensure that each resource experiences first hand that mode of operation, and enjoys the outcome. This can be accomplished by using the freed-up time to complete the preparations on the running projects. The things that are missing are usually things for which there is some difficulty to complete. Therefore, if given the option, resources working on preparations would prefer to focus on preparing new projects about to be released rather than relentlessly chasing the preparations gaps on open projects.

Not only is it possible to wait long enough to start projects with full kit and yet deliver on time, if we quickly use our newly available resources to complete full kit (first on running projects then on frozen projects), we get at least two additional benefits. The first is that we overcome the human tendency when overloaded to do the easiest things first. This is an improvement over the current situation, where because of the work involved in full kitting, projects held for various shortages tend to be neglected in lieu of new projects – thus creating even greater delays for open projects, as the new projects themselves eventually become stuck, and hence, are at least partially neglected for even more "urgent" projects.

The other benefit is that the organization immediately gets to experience the benefits of "life with full kitting," an environment far less chaotic and far more stress-free. This not only makes the change easier to accept, but much easier to maintain, since the "old way" quickly becomes undesirable in comparison. The key, of course, is that we move immediately after freeze to accomplish full kitting.

This parallel assumption also contains a gem that may not be obvious upon first glance. The phrase, "if given the option, resources working on preparations would prefer to focus on preparing new projects about to be released rather than relentlessly chasing the preparations gaps on open projects," implies heavily that the ideal state of the company is to *relentlessly* chase the preparations gaps on open projects. Dr. Goldratt used to say that every word in the S&T trees are important and that the words were chosen carefully.

So let's take a closer look at this word, "relentlessly." Here is the dictionary definition of "relentless." As an adjective, relentlessly means someone or something:
...that does not relent; unyieldingly severe, strict, or harsh; unrelenting: a relentless enemy.

It is clear, then, that the ideal posture of full kitting for the organization is "*unyieldingly* severe, strict, or harsh." It means that we pursue full kit as if our lives depended on it. And if we want our organizations to survive, grow, provide job security and be sustainable many years into the future – if we are talking about our organizational lives – they very well may.

Tactics

A Full-Kit manager is appointed. The relevant resources are instructed to complete the preparation steps first for the running - not frozen - projects. Then to complete the preparations for frozen projects. Only when (most of) the above is done they are guided to work on the preparations for the new projects waiting to be released. They always follow the projects priority.

It is often said that the job of a Chief Financial Offer in a company is to say "no." Since the CFO has primary fiduciary responsibility for the company, he or she must be the guardian at the gates of the company's finances. Therefore all major purchases and investments must include the approval of the CFO. The CFO will demand a rock-solid business case before making any approval. And even then, the CFO's signature alone is often not enough. Large or strategic financial transactions will also require the signatures of other top company officers, such as the President or CEO.

This is a perfect analogy to the Full Kit Manager. It is his or her job to say "no" when people request release of projects (or legs of projects) when they do not yet have full kit. The Full Kit Manager has primary responsibility for the uninterrupted flow of the company's precious throughput-producing projects. In this regard, he might just be the CFO's best friend! Like the CFO, the Full Kit Manager should demand a rock-solid "business case" as to why any project should be released without full kit. And even with his recommendation, the signature of one or more other top officers should be required before granting a request that has the potential of creating severe negative effects in the company's operations.

In another sense, preparation, or full kit, is the responsibility of everyone in the organization. Every employee, whether he or she be an executive, project manager, production worker, task manager (supervisor), etc. has a responsibility to make sure they are prepared to execute their responsibilities when called upon. For example, task managers should ensure that the workers to which they will assign tasks have what they need to perform the tasks –tools, certifications, consumables, etc. without delay or defect. Therefore task managers should be looking forward for a reasonable period of time, making sure preparations for upcoming tasks are complete. We call this task-level full kit.

However in the project-oriented organization, full kit has an even more specific meaning. This is full kit at the project level. Projects are the life-blood of the company. Projects employ people, and put food on the tables of many families. It is therefore critical that projects are executed in the most effective way possible. The stakes are high – effective projects sustain the company and help it grow. Therefore, the full kit process is one of the most important processes for the organization. It is so important that it should not be left to chance, or even to the "good intentions" of employees. The best way to have effective full-kitting of projects is to have the company manage it officially – with a specific function with specific authority – the Full Kit Manager. Given the importance of this position, the Full Kit Manager deserves the respect and support of everyone in the organization. Therefore a current project manager should never be the Full Kit Manager. In fact, he or she should be higher in the organization than project managers. And there should never be more than one official Full Kit Manager. Make no mistake – this is a management position – one that requires superior knowledge and experience with the products and processes of the company.

Flow is the number one consideration of the Full Kit Manager. Flow is the reason a Full Kit Manager exists – the reason he or she is necessary. The flow of projects through the company is both a function (role) and primary responsibility of the Full Kit Manager. The Full Kit Manager is a steward for the holistic flow of the organization. However, the responsibilities of the Full Kit Manager extend much further than this. Not only is he or she the gatekeeper and flow expert in the company, but the position should enable him or her to identify the weakest links *in the supply chain* for further improvements of flow. The Full Kit Manager's responsibility and authority should extend as deeply as negotiating change in contractual relations with suppliers, overriding local optima problems as such saving costs on shipments, short-cutting prolonged supplier price negotiations that lose weeks of buffer, etc. So while flow is the number one consideration of the Full Kit Manager, it is the *flow of the entire supply chain* that is truly his or her number one consideration.

NOTE: Close examination of the cause and effect of supplier late deliveries has shown that sometimes a high percentage of late deliveries are actually the fault of the *purchasing* organization, not the supplier. The reasons for this include unclear or missing specifications and drawings, engineering change requests subsequent to order, failure to make previous payments to the supplier on time (or at all), requests for shipping efficiencies (waiting to ship a large batch of parts in order to save shipping costs), and many more. Therefore part of the Full Kit Manager's job also includes insuring that we are not our own worst enemy (see Chapter 14 for more information).

Unlike project managers, who are concerned mainly with the health and delivery of their own assigned projects, the Full Kit Manager is one of the key people directly responsible for the overall, holistic flow of the organization. He or she must have sufficient experience to understand, at a basic level at minimum, all aspects of the organization's projects – work content, materials, key resources, etc. He must have sufficient authority to:

1) mediate and decide for the organization in conflicts for resources between project managers,
2) hold projects out of execution, and to allow them into execution.

To be able to do these things effectively, in addition to possessing the necessary experience and competence to perform the required duties, he/she must have the full backing and support of senior management.

In organizations where there are few and/or very small projects, the Full Kit Manager may be a part-time position. In organizations where there are numerous projects, and/or the projects are large, this position will be full time. In very large organizations, with multiple large projects, the Full Kit Manager may even require a staff to support the meeting of his responsibilities and to ensure the holistic flow of the organization. In an organization with many locations and divisions, the Company Full Kit Manager may even oversee Full Kit Managers by division (or site). The ultimate objective of the Full Kit Manager is to create a *culture* of Full Kit across the company. Evidence of such a culture would be, as the Strategy says, "Resources and project leaders are used to working on projects whose preparations are (almost) fully completed."[34]

Part of the Full Kit Manager's responsibility is to oversee the development of the infrastructure, policies and procedures for creating and maintaining the full kitting department. It is his or her responsibility to ensure that checklists and templates are documented for projects and the templates are correctly modified and customized for new projects.

It is the Full Kit Manager's responsibility[35] to ensure that full kitting is done in priority sequence: open projects first, frozen projects second, and to-be-released projects third, according to the start dates of the "legs" of the projects. The Full Kit Manager monitors the "filling" of the kits for the legs, and takes appropriate actions (with vendors, other departments, etc.) to ensure that kits are filled in the fastest, most effective manner.

---

[34] In engineering and procurement (activities sometimes referred to as "pre-manufacturing"), Full Kit is just as important as it is in the manufacturing segment of the company.

[35] Because someone has responsibility does not mean that person has to do all the work. Remember, Full Kit Manager is a management position.

When the full kit of a leg is ready, and its time has come to start according to the schedule, the leg can be released into work.

It is also the Full Kit Manager's responsibility to work with resource managers and identify critical resources (including critical skills)[36] – those which if not available will absolutely delay the start of a leg, and work diligently and creatively with buyers, subcontractors, managers, etc. to make sure these critical resources are available for the start of the leg. Obviously, when the leg of the project is the Critical Chain, this responsibility is exponentially greater.

Other than the critical resources, the organization should determine how much of the full kit is necessary to allow the start of a leg (some experts recommend the minimum should be 95%).[37] Assuming all critical resources are available, the Full Kit Manager will not release a leg for production if he does not have on hand at least the specified percentage of all required inputs for the leg. Even the Full Kit Manager cannot violate this number on his own authority. This is an unusual situation which would require senior management approval. The purpose of this is to ensure that exceptions to the rule are not growing over time, returning the organization to chaos.

Was the step correctly implemented?

Here you will be verifying that full kit has indeed begun in priority order, first on open projects (as far to completion as possible) then on frozen projects and finally on to-be-released projects. You will verify the Full Kit Manager has been appointed, is at the appropriate management level, has the respect and authority required, and is supervising the full kitting process. Finally, you will confirm the documentation and adherence to rules and signature authority required for exceptions.

---

[36] Some of the best organizations include a verification of available optimal resources as part of their full kit.

[37] Personally, I find such a number to be arbitrary and dangerous, and do not recommend such a policy. Of the 5% of the full kit that might be missing, exactly *what* is missing and *when* will it be needed? Is it something that will bring work on the project to a screeching halt? Is it something needed in the first week of the project? Is it something coming from an undependable supplier? Please be very careful with your definitions and other criteria when considering the adoption of such a percentage.

Were the expected effects realized?

The expected results of this step are that open projects are able to move with un-interrupted flow, that frozen projects can be released when signaled by a Virtual Drum completion (because they are not lacking full kit), and that people are starting to acknowledge the benefits of operating in an atmosphere of being "used to" full kitting.

What mechanism has been put in place to ensure compliance with the step over time?

Before moving on to the next step, measurements should be in place to ensure projects continue to be kitted in priority order by project and by leg. You should also be watching for negative trends: are we releasing more projects/legs without full kit? Are we granting more exceptions? Given the damage such trends could lead to, it is vital that if recognized, these trends are dealt with immediately. It is also very valuable for the Full Kit Manager to monitor delay reasons from completed projects relating to shortfalls in the full kit. What percentage of delays could have been avoided by more diligent and effective adherence to our full kit policies?

Chapter 13 – 5.112.2 Defining Preparations

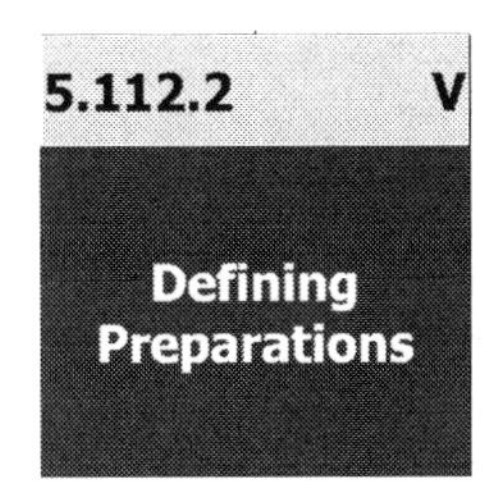

The dictionary defines preparation as "the act or process of preparing." Therefore, when we "define preparations" we are creating a process, or documenting the way we prepare to execute projects. We are making the process of preparing our company policy. The result of these preparations is something known as the "Full Kit." The terms are used in CCPM interchangeably, but really the "Full Kit" is the output, or result, of preparations.

In accordance with the one-step-at-a-time per functional manager rule, once full kit has been largely accomplished on open and frozen projects, and full kitting for to-be-released projects has commenced, the Full Kit Manager can turn his or her attention to defining preparations (the process) and establishing the full-kit infrastructure.

To create the process, build the infrastructure, establish a culture of full kitting, and to sustain this culture into the future, we will need a solid set of strategy and tactics.

Necessary Assumptions

The permission to work on preparations on frozen projects (and even projects that are not yet released) might be misused by eager project leaders to pressure resources to work on more than just preparations on frozen projects – flooding back the system with work.

Preparations. What are they? Technically, one could say that everything up to shipping is preparation for delivery. Without a clear definition of what preparations are and what they are not, people can find many ways of working on projects while calling it "preparation." Col. Steve Foreman, the retired Commanding Officer of the U.S. Marine Corps Maintenance Center introduced in Chapter 1, once said that he could take a vehicle that was suspended from work, park it against the back fence, station a Marine with an M-16 rifle on it with strict instructions not to let anyone touch it, and somehow project work would still get done on it. So unless we fully understand and define what constitutes preparations (and just as importantly what does not), unnecessary work can overload our resources, impede flow and lengthen our lead times.[38]

So where do we draw the line? What constitutes preparations in the sense of the S&T tree, and what does not? How do we keep "eager" people (most likely with a desire to do good for the company) from doing work that in reality is damaging to flow and the health of our systems? It certainly appears that a carefully considered strategy is necessary.

Strategy

The permission (or even demand) to work on preparations does not violate the freeze and/or controlled release intentions.

It may seem strange that this strategy is based not on a rule, calculation or formula, but instead on the basis of being consistent with the *intentions* of some of our earlier tactics. When considered in this way, we have an extra dimension of meaning, and therefore we get increased clarity on the strategy. So just what was *intended* by the tactics of freeze and controlled release? Focus!

First, the intention of the freeze step was to *maximize flow* by lowering Work-In-Progress – like reducing the number of cars from a jammed freeway. Next, the intention of controlled release is to *sustain*

[38] Another problem related to work being done early and overloading the system is that cannibalization, i.e. the stealing of parts from one project to fill a shortage on another, is hidden much more easily.

maximized flow by metering the release of new projects to be in line with the lowered WIP levels. In addition but not less important is the fact that full kitting (the result of the preparations policy) minimizes preventable disruptions in flow by assuring the project has everything it needs to proceed. It is clear that **flow is the number one consideration** (or intention) of each of these steps.

When the pressure to release projects or legs without full kit inevitably comes, many "good" reasons are cited. Unfortunately, most of these good reasons fail to consider their impact on flow, and if continually allowed these exceptions will deteriorate the system to the point where everything hoped for from the exceptions will be impossible to achieve anyway. Therefore the negative impact of releasing without full kit becomes greater than the hoped-for positives. The interests of flow must always trump other reasons, and every request for an exception to the full kit rule must be considered in light of the *intentions* of the freeze and controlled release steps.

When considered in this light, the ability to say "no" is enhanced, and the system is better protected.

Parallel Assumptions

There is a good intuitive understanding of which activities are regarded as preparations and which are not. In most multi-project environments there is no formal definition of which activities are entitled preparations and which are not.

Here we have two parallel assumptions. The first says that if we are honest with ourselves we really do know intuitively the distinction between preparations and project work. This means we have in our collective knowledge everything we need to know to separate the two, follow the *intentions* of freezing and controlled release, and avoid later disruptions.

The second assumption says that in most project environments, although this knowledge exists, it is not documented or followed in any systematic way. It won't take long to validate this in your own environment. The good news is that this gives us an opportunity to really improve the sustainability of keeping Work-In-Progress at a level that maximizes flow, as well as to cut to a manageable level the frequent but easily preventable flow disruptions and delay events. Given the apparent validity of these two assumptions, it is easy to arrive at a set of constructive tactics.

Tactics

The activities which should be titled preparations are officially defined as such. The company takes the actions to ensure that resources (those conducting the preparations and project managers of frozen and unreleased projects) are guided and monitored to work only on the preparation activities as defined.

Since tactics are the action steps employed in order to achieve our strategy, it is here that we will delineate the actions necessary and sufficient for not only defining preparations, but for putting in place a permanent Full Kitting operation. This involves more than just creating full-kit checklists, although this is a crucial component of the process. It involves building the infrastructure for permanent and sustainable project full-kitting and putting in place the mechanism to ensure people are "guided and monitored to work only on the preparation activities as defined." In a sense, this is the "Level 6" S&T activity that takes place under many Level 5 steps, but are usually not documented as such because they are self-apparent. However, in the future it may be deemed wise to add Level 6 entities for this and perhaps other Level 5 entities.

It was mentioned in previous chapters that the identification and training of permanent full-kitting resources should begin in accordance with the Freeze step. It would be natural for many of the permanent full-kitting personnel to come from the Procurement and Materials departments, as there are many similarities with the work they are presently doing. However given the importance of full kitting, the search for the right people should coincide with the Freeze step (or even begin earlier and extend across the entire company).

How many full kitting resources are needed? A great TOC answer is "enough." Enough for what? Enough to sustain a culture of full kitting both in theory and in reality, growing as necessary as the company and workload grow. As was stated earlier, these people should be "relentless" at providing full kits for all legs of all projects at the proper time. It may be natural that these resources report to the Full Kit Manager, but this does not mean that the Full Kit Manager should necessarily be burdened with the other duties of the materials and procurement departments, unless it is a very small company.

After full kits have been provided to in-progress and frozen projects, beginning with the to-be-released projects – thorough documentation in the form of checklists should be created for all legs and all projects. These checklists should always be retained in case another project like this comes along in the future. However in the cases where identical or similar projects are *expected* to be repeated, these should officially be designated as templates and kept on hand to be reused or modified as necessary when the situation arises.

In addition to personnel and checklists, for environments involving physical hardware a physical staging area for full kits may be necessary. This staging area can take many forms, but it should provide for clear visual identification of missing items, the filling of which should be pursued *relentlessly*. At the Albany, Georgia Marine Corps Maintenance Center, the staging area was a warehouse with painted "parking stalls" on the floor for pallets upon which the full kit was constructed. Portable signs indicating the expected date of release were placed clearly behind the pallets, and when the kit was verified as full and the release date arrived, a forklift picked up the pallet and transported it to the work location. Another form of a staging area is shown in the pictures below.

Figure 18. Full Kit staging area for large industrial company in India

Different environments may require unique full kitting infrastructure. As of this writing, a couple of the better CCPM software packages include full kitting capability, for example electronic copies of the full kit checklists. But outstanding full kitting can easily be done with a lower-tech approach as well. The organization should spend the appropriate amount of time determining and building the necessary elements to always ensure full kit for (almost) all projects.

Was the step correctly implemented?

The step is correctly implemented if you can verify that every leg of every project has properly detailed full kit documentation, the necessary infrastructure exists to support a culture of full kitting, and in actuality all legs of all projects do not start without full kit, except in rare, top management-authorized instances. Since a huge percentage of delays in projects can be traced to lack of full kitting, the organization cannot afford to be distracted from or give a half-hearted effort to full kitting. It is a win for the company, a win for the workers, and a win for the customers.

Were the expected effects realized?

The expected effects of the preparations definition steps should be self-evident: projects are rarely (meaning almost never) delayed as a result of missing parts, tools, documentation, etc.[39] This translates to increased flow, earlier project completions and reduced lead time. When the organization has greatly decreased task switching and ensured full kit by leg for their projects, their improvement will be dramatic and apparent to all.

What mechanism has been put in place to ensure compliance with the step over time?

Measurements can help us ensure that "resources (those conducting the preparations and project managers of frozen and unreleased projects) are guided and monitored to work only on the preparation activities as defined." As always, your organization should have serious discussions about what to measure (and what not to measure). Of course, any measurement must be in compliance with flow being the number one consideration. In other words, any measurement that impedes flow, such as measurements relating to local efficiency policies – good for the local area but problematic elsewhere (e.g. batching to save setups), should be discarded in favor of those which enhance flow (e.g. small batches or one-piece flow).

In the case of defining preparations, suggestions for measurement include:

1) How often was the full kit checklist unprepared/unavailable?
2) How often did the full kit checklist contain omissions or improper additions?
3) How often were exceptions requested (even if not granted) due to shortages?
4) How often was a project or leg unable to start due to a less than full kit?

[39] You might want to verify this by measuring the number of delays caused by missing inputs both before and after implementation and analyzing the magnitude of improvement.

One helpful idea is to create a "full kit buffer." This means choosing a timeframe, for example two weeks before a leg starts, when full kit should be complete, and use green/yellow/red buffer management to monitor and manage your full kitting activity. Then you can also measure how often your full kit buffer was red and look for negative trends. You can also use the POOGI[40] process to identify and track related delay reasons (see Chapter 30).

[40] Process of On-going Improvement.

Chapter 14 – 5.112.3 Worried Clients

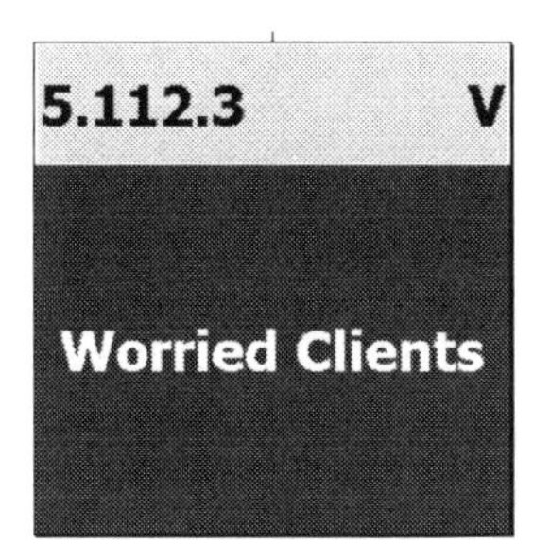

Prior to his passing in 2011, Dr. Eli Goldratt was working on a new book – the unpublished "Science of Management." The last thing he wrote was two pages on the subject of fear – the thing we now believe to be the main constraint of the human being. His premise was that fear - fear of complexity, fear of the unknown, and fear that conflict will lead us into a tug-of-war – can cause us to fail to do the right thing and make decisions that are not in our best interests. Although the "Worried Clients" step is in itself rather narrow and specific, it presents an excellent opportunity to expand its logic to a much wider and more general sphere of daily relationships with customers, suppliers, and even others within our organizations. So with your permission I will take that liberty. Therefore, in addition to dealing with worried clients when new projects are frozen, this chapter is also about the subject of fear. Furthermore, it is about communication. Communicating with our customers (and our suppliers) often reveals that our fears are unfounded. Communication helps us to make better decisions and pre-empt conflicts.

The paradigm shifts involved with CCPM implementation can generate fear if they are not properly understood and communicated, both internally and externally. These paradigm shifts are necessary for creating ever-flourishing companies. The results can be unbelievably rewarding, both in financial and human terms. But the freezing and full kitting steps, in delaying the start (or continuation) of projects, can excite fears that our customers will not understand our actions and will punish us for doing the right thing. It would be a shame if we forfeited the reward for doing the right things because of our fear. The premise of this chapter is that these fears can be evaporated, and we can proceed with the customer's approval. A good place to begin is with strategy and tactics for dealing with worried clients.

Necessary Assumptions

In some environments, delaying the release of projects might cause the exposure that clients (seeing that no work had started on their project and concluding that their project will not be ready on time) might transfer their project to a competitor.

Although we may feel fear of various levels of punishment from customers for delaying projects, the ultimate fear is that by doing so we may lose their business entirely. If we truly believe this may happen, it is a powerful enemy to us for doing the right thing. But the damage done by making an unwise decision to start early (the wrong thing) based on this fear is very real, and we need to do what is necessary to make sure that the customer is satisfied – while at the same time we are able to take the proper actions.

Strategy

The threat of losing projects due to a late start is alleviated.

The strategy we need to pursue is one where we do not have the fear of losing clients. We want our customers to have confidence in us that we know what we are doing. This means we cannot generate fear in them – fear that their projects won't be delivered on time.

Parallel Assumptions

Clients of projects are fully aware that proper preparation is essential to shrink the lead time of a project. In most environments some preparations involve the client's efforts. In all environments it behooves the Company to regularly report to the client about progress/difficulties

in preparations. Therefore the full exposure exists only in the window of time when preparations are not allowed to be done on new projects (the time until the freed-up resources complete the gaps on open and frozen projects).

Parallel assumptions are statements about our reality, guiding us from our strategy to the best tactic for accomplishing the strategy. In this case we find a wealth of insight in a couple of sentences. The first says that our customers are fully aware of the importance of preparations in shortening lead times, i.e. getting their projects delivered as fast as possible. This is especially interesting because our actions (and lack of certain actions) seem to indicate we don't actually believe this. Nevertheless there is a tremendous amount of evidence available that this is true.

In fact, most of the customers of project companies are project companies themselves. For example, when General Electric delivers a jet engine (a project) to Boeing, GE is delivering their project to another project company (building an airplane is a project). Project companies know the importance of preparations. It is almost a certainty that the calls for preparations in the internal operations of the customer company are just as loud as they are in the supplier company (see the Necessary Assumption in 5.112.1, Chapter 12).

The implication here is that if we are asked by our customers why we have frozen their projects, we can and should inform them about the actions we are taking to become a permanently reliable supplier (our TOC/CCPM journey and all its steps including freezing and full kitting). In addition, we should not let our fear of the customer's response stop us from taking these actions in the first place. Since customer companies are also likely to be project companies, they will (to a large degree) understand and support us when the focus on preparations is strong and clear. I have personally witnessed this on a number of occasions, even with companies in the Fortune 500, companies that people fear "would never listen to little, old us." In reality, they understand and appreciate our efforts. They may still be a bit wary and demand that we follow through with results, but they almost assuredly will not take their business elsewhere. I have never seen it happen.

Now let's expand this thought. This step is technically about' the temporary window of "exposure" to our customers, when we are NOT performing preparations on new projects because we are first completing preparations on open and frozen projects. But the same concept applies to the new paradigm when we are no longer starting all projects immediately, but rather we stagger their releases according to completions on the Virtual Drum. What if the customer calls and asks us for status on their project, expecting us to have started it?

The WRONG answer is, "Oh, we haven't started it yet." This answer implies we are not working on their project. But nothing could be further from the truth. We are indeed working on their project – on preparations – ensuring that full kit is available when the legs begin so that there are no delays during execution and we finish on time. So the right answer is "We are working on preparations and completing full kit so we can ensure on-time delivery." And said with sincere resolve, that has a positively calming and confidence-inducing effect on the customer.

The second assumption states, "In most environments some preparations involve the client's efforts." This can be in the form of engineering, approvals, inspection, and sometimes even materials. Once again, communication plays a major role in helping the customer help himself when it comes to on-time delivery. Especially in the case where your customer is a big company, he is constantly bombarded with questions and requests – not just from you, but from all his suppliers. To the customer's ear, everything sounds urgent. As in your previous existence, your customer is likely multi-tasking and thus overloaded. He also knows that even though everything sounds urgent, some of the pleas for help are actually not urgent at all. He knows that some of the "urgent" calls are just people trying to check things off their lists. But how can he know the truly important from the not so important? In the traditional system, he can't, and when everything sounds urgent, nothing is urgent. Since your customer does not know what is truly important and what is not, he will create his own priority list. Often service is given not by true urgency, but by who is yelling the loudest. Good luck on getting the fulfillment of your need on time.

In your brave new world, you will now have the tools to show the customer the true impact that a delay – by him – will have on the delivery of his project. You can show the customer not only the point of the delay on the Critical Chain, but also its impact on the project buffer. If you educate your customer on how to understand your system, it will be clear to him you are only calling when the need is genuine. In the past, it sometimes might have seemed the customer didn't care about your problems. In reality, he did not know how to sift the truly urgent requests out of the noise. Now, and I can verify this through my own experience, he will see you as trustworthy and will be happy to help you with your request.

The next assumption itself expands this idea of communication to all of your business relationships: "In all environments it behooves the Company to regularly report to the client about progress/difficulties in preparations." Therefore this is not only applicable for the "Worried Clients" step, but for your day-to-day operations as well. It also applies when the customer is not the cause of the delay in your preparations. When the customer is aware of your difficulty with a third party, he may be able to help, or at least be better able to understand the nature of the delay. Communication is the key.

The last assumption brings us back to the subject of Worried Clients. Everything said so far is based on communication reducing or eliminating fear – our fear of losing business and the customer's fear of late delivery. Communication usually works wonders – we often find out that not only were our fears unfounded, but by not communicating we can actually be harming ourselves. However, although this is generally proven out in reality, there may be a small number of cases when the customer cannot or does not *want* to understand. In spite of your explanation and honesty, the customer might threaten you anyway (you never know how he might be being measured). In such cases, the period of danger is still relatively limited: "the full exposure exists only in the window of time when preparations are not allowed to be done on new projects." When a culture of full kit has been established in the organization and the company is truly working diligently on preparations, it can honestly be said we are making progress on projects, even when we have not yet begun actual project work. But it is crucial that we take preparations very seriously, and are closing gaps *relentlessly*.

Tactics

The Company relentlessly completes all preparations (closes the gaps) on running and frozen projects. Note: At that stage preparations are done on new projects and the exposure is drastically reduced. Once the Reliability offer is properly launched the above threat is completely removed.

There's that word again – *relentlessly.* We attack preparations relentlessly for two reasons: First, we need to complete preparations on open and frozen projects as fast as possible. This reduces the window of "exposure," and helps us avoid future potential delays. Second and perhaps more importantly, it helps us to develop the culture of full kitting. There is no explicit or implicit idea here that we will cease to work "relentlessly" once the window of exposure has passed. We've only just begun.

The last sentence in the tactic is a reference to the Reliability Decisive Competitive Edge in entity 2.1 of the S&T tree. Although this book is not necessarily dealing with a full "Viable Vision" or Reliability DCE, I encourage you to read the entire S&T tree and decide if these things are right for you.

When you are your own worst enemy

Finally, I would like to introduce a subject that is not explicitly found in this step. Nevertheless, given the subject of communication, this is an excellent place to talk about it. It relates directly with full kitting and applies when you – not the customer or the supplier – are actually the cause of the delay in completing preparations.

It is very common when projects are scheduled to start (and even more common when you start everything early) that certain key materials or parts are missing. Another common aspect of this situation is that even when projects you started early eventually become late, the finger is pointed at the supplier as the culprit. When our customer demands to know why we are running late, we are quick to shift the blame to our suppliers. But shifting blame does not address a huge problem – one where we consume a large amount of buffer before we even begin production. When this is a chronic situation it drives the production department crazy and launches a verbal war between them and the upstream functions. But the truth is, before we shift blame, we should look in the mirror. At one large multinational corporation in which I implemented CCPM, it was discovered that the cause of up to 80%(!) of the missing components was the fault of the company I was working with.

How is this possible? Many reasons – including:

1) Ordering the parts late (often after prolonged negotiations for a lower price)
2) Sending specifications to the supplier late
3) Asking for engineering changes after ordering, and even
4) Failing to pay the supplier for previous orders (and this was a big, rich company)

The Full Kit Manager and his or her team should check closely whether any of these situations exist within their projects, by priority order, of course. One way to make this easier is to model the procurement process properly in the project network (see Chapter 16). Very often, if procurement exists at all in the project network, it is represented by one or two tasks, for example "Place Order" and "Order Lead Time." Unfortunately there is very little we can learn from such a breakdown. Many times we call this level of detail a "black box," because what goes on inside is a mystery. The problem is in translating what is happening in procurement to its impact on the project schedule. In a case like this, the "Order Lead Time" task may be weeks or months long. This makes it very difficult to update remaining duration of the task with much accuracy (see Chapter 21). If the task is 15 weeks (75 days) long, is reporting after 1 day that the remaining duration is 74 days relevant? How do we know it's 74 days?

Let's open the black box and look inside. The details may vary from company to company, but the types of activities that can exist within the black box are, "Create Engineering/Specifications," "Select supplier," "Place purchase order," "Deliver specifications," "Manufacturing/Fabrication," "Inspection/Approval," "Transportation," "Receiving Inspection," etc. By using this level of resolution, we have much better visibility on how the procurement process is impacting the project, because delays in any of these steps will consume the project or feeding buffers. For example, if the estimated duration for "Select supplier" is 5 days, we will be able to see early-on whether we are wasting weeks of the buffer negotiating over lower component prices – negotiating over pennies in what might be a multi-million dollar project. Likewise, for "Create Engineering/Specifications," we can see if we are consuming buffer waiting for engineering for a task required on the Critical Chain. Furthermore, the task managers for Engineering and Procurement can see the relative priorities of these tasks across all projects.[41]

As you can see, the information you can glean from this level of resolution is much greater and can help avoid late starts and early-project buffer consumption. This is one example of what is meant by "relentlessly completing preparations." Therefore since this part of the conversation does not specifically relate to "Worried Clients," it certainly does directly relate to communication and preparations.

Was the step correctly implemented?

In determining how to answer the Three Questions for this step, we should consider both the situation specific to "Worried Clients" during the window of exposure, and also consider the expanded overall communication and preparations issues. First, you will want to check whether we have honestly communicated with the client the reason for their projects being placed on hold (the reason is long-term reliability, but the last elephant will still go through the door earlier than if we did not implement!) We also need to verify that we were *relentlessly* completing preparations – closing the gaps – in open and frozen projects so we can quickly get to preparations on to-be-released projects.

For the expanded aspects of this step, you should confirm that communication is occurring both directions in the supply chain (both customers and suppliers), and that the Full Kit Manager and his team are succeeding in making the relentless pursuit of full kit the new company culture.

Were the expected effects realized?

In the case of Worried Clients during the window of exposure, the expected results are that clients are at least somewhat comforted by the fact that you are expending all efforts to become a permanently reliable

---

[41] This does not mean we need to model every component on the bill of materials. But we might want to model long-lead items and critical items that tend to be problematic, which can delay the start of legs/projects.

supplier. In the expanded case (and over the longer term), the expected results are greatly reduced delays early in projects, faster full kitting, and increased harmony between the Production Department and upstream functions.

What mechanism has been put in place to ensure compliance with the step over time?

In this unique case, a mechanism is not required for the single-occurrence window of exposure. However a mechanism or mechanisms are very appropriate for the expanded case.

You should devote some time to discussing the type(s) of mechanism that are best for your environment. Some ideas include: For communicating with customers, we might measure their satisfaction with our procurement system, things like whether we are delivering specifications on time, whether we are creating chaos for them with our change requests, and whether of our payments for their services are prompt and complete. For communicating with suppliers, we might first, by project, ensure the granularity of our procurement steps is sufficient, and we might include a full kit buffer that alerts us when any task in the procurement phase is endangering the start of any leg.

Now we have lowered the load on the system, accelerated projects and created a robust culture of full kitting. Once we have confirmed the preliminaries are complete, we can move to the next phase of our implementation – we are ready for “the main event.”

Section Four: Critical Chain

Flow is the number one consideration

Chapter 15 – 4.11.3 Critical Chain Planning & Buffering

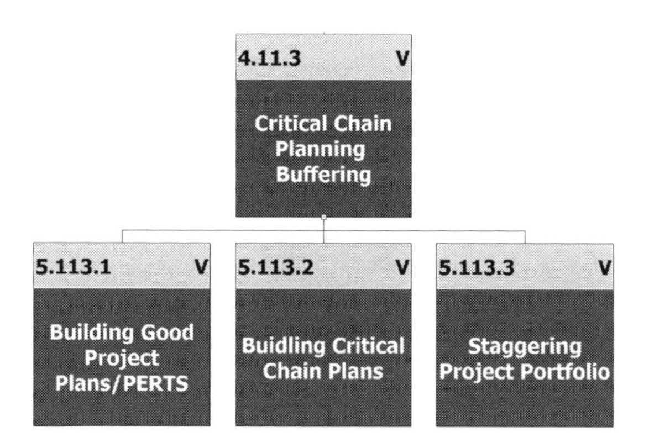

At long last we reach the point in the Handbook where we begin to discuss the main subject, Critical Chain Project Management. But really the activity described in these chapters should begin soon after freezing, keeping in mind the one-step-at-a-time rule for an individual functional manager. To get a general idea of when these activities can begin, please refer to Figure 17 in Chapter 10. In order to successfully accomplish the strategy of 3.1.1, "Meeting ALL Project Promises," we must attain each of the strategies in 4.11.1 through 4.11.6. 4.11.1, "Reducing Bad Multi-tasking, WIP," 4.11.2, "Full Kitting," and this section, 4.11.3, are foundational for 4.11.4, "Managing Execution," and beyond. It is crucial that before we actually begin executing the complete CCPM system that these three sections are completed successfully.

The steps that we implement prior to launching CCPM are extremely powerful, and even if they are the only steps taken, the organization will experience tremendous improvement in project operations. However to become a world-class projects company – consistently fast, reliable, and ever-flourishing, we also need a way to deal with the day-to-day variations that occur during normal execution of the various tasks in a project. Finally, we need to have a robust mechanism for making commitments. The steps in this section, 4.11.3 Critical Chain Planning and Buffering are meant to accomplish this. If traditional project management is an acorn, and the steps leading to Critical Chain are an oak tree, Critical Chain, properly implemented, is a forest.

Necessary Assumptions

Contrary to the common belief, safety embedded at the task level prolongs the project without providing sufficient safety to the project completion. Contrary to the common belief, having detailed visibility (having too detailed a PERT network) almost guarantees that control will be lost.

Anyone who has been around projects very long knows that due to variability and its big brother "Murphy," we need to have safety in our projects. But what is less understood is that safety at the task level actually prolongs – rather than protects our projects. This becomes clear when you understand the human behaviors involved around the subject of estimating task durations. These behaviors include Student Syndrome, Parkinson's Law, and more, and are described clearly in Eli Goldratt's book "Critical Chain," and many other books and papers on Critical Chain theory.

Even less well-understood is the second assumption, which while implying that PERT (Program [or Project] Evaluation and Review Technique) charts are a valuable tool for the execution of projects at the individual project level, they become unmanageable if they are too detailed.[42] It seems quite clear then, that if these two identified common beliefs are actually wrong, we need a strategy that helps us manage projects more effectively.

[42] Many companies do not routinely use PERT Charts. The use of GANTT charts alone is far more common. But the point on the level of detail still applies. It is easier to see detail-overload in a PERT because everything is displayed. But in a GANTT chart, many details can be "hidden" beneath summary bars, so be careful. In TOC, we recommend using both types of chart, doing the actual project network construction in PERT form, thereby having better visibility on the level of detail in our projects.

Strategy

Flow is the number one consideration (it is not important to finish each task on time, it is essential to finish each project on time).

Flow is the number one consideration. Where have we heard that before? This strategy introduces yet another paradigm shift. In a world where the fear of complexity[43] drives us to break things down into smaller and smaller elements in hope of making them more manageable, this strategy reminds us of what is truly important – finishing our projects on time.

In traditional project management, manageability is assumed to mean the ability to protect the start and finish dates of each task. It is believed that this ability improves with greater granularity. However, this has not been proven. Furthermore, the effort to improve manageability through more and more granularity increases exponentially, so that any incremental increase in control is not justified by the required investment. As the effort required becomes overwhelming, we push ourselves into chronic task switching to try to complete all tasks on time – some of which can actually afford delay without endangering the delivery dates.[44]

But if it is not important that individual tasks finish on time, what should our tactics be for finishing our projects on time? For that, we look to the Parallel Assumptions.

Parallel Assumptions

The bigger the uncertainty, the bigger the safety embedded in the task's time estimates. In the vast majority of project environments safety is at least half of the time estimate. Shifting the safeties from the tasks to the end of their respective task sequences (paths) not only places the safety in the place where it should be but also requires much less safety than the sum of safeties removed from the tasks. This requires that resources

---

[43] Or the desire to "be in control" in the face of dynamic complexity.

[44] This also makes earned-value management systems problematic in terms of finishing on time – we are incentivized to complete tasks which might not be critical at the moment before difficult ones, in order to gain present earned value. Therefore flow ceases to be our number one consideration.

will no longer be judged by meeting their time estimates. Critical Chain Project Management (CCPM) planning methodology provides a proper guide for where and how much safety should be inserted in project planning. To get excellent control, it behooves keeping the number of tasks in the PERT network to less than 300 (for huge projects zooming might be needed). Using templates (when applicable) significantly reduces the planning time and reduces unneeded variations.

Inherent simplicity. It was one of Eli Goldratt's Four Pillars of TOC. There may be no better place to highlight the concept of inherent simplicity than in the subject of task durations and safety. The parallel assumptions here provide us with a wealth of information. Unfortunately, we are so accustomed to looking for complex solutions to complicated situations that we easily miss the inherent simplicity embodied here. This applies even to veteran TOC practitioners.

Let's take a look at some of the statements above, and try to understand them from the standpoint of inherent simplicity:

1) In the vast majority of project environments safety is at least half of the time estimate.
2) Shifting the safeties from the tasks to the end of their respective task sequences (paths) not only places the safety in the place where it should be but also requires much less safety than the sum of safeties removed from the tasks.
3) This requires that resources will no longer be judged by meeting their time estimates.

Read each of these assumptions, one at a time, and understand how very simple and straightforward they are. Then try to realize, within this simplicity, how profound they are. First, we are told that outside of exceptional cases, task durations quoted in the traditional manner contain *at least* 50% safety. This means that if we add the safety from all tasks together, currently our *projects* contain at least 50% safety!

The second assumption, that a) the best location for placing safety is at the end of each path (or chain of tasks), and b) that because not every task will have a problem not so much overall safety is needed, opens up a new dimension for us. A common misconception is that individual tasks have buffer "allowances." If those who haven't made the full paradigm shift from managing projects at the task level to managing at the project level are asked the question, "How much buffer does a 10-day task have?" the common answer is "5 days." But this is not the correct answer. Buffers are like insurance policies – you use what you need. The applicable chain's entire buffer is available to each and every task. This is easier to comprehend if the safety is aggregated. You might think of a buffer as being like a cloud, moving from each completed task to the next task and "hovering" over it to be used as needed. *Buffers are there to be consumed* (but never to be wasted).

The third assumption, that "This requires that resources will no longer be judged by meeting their time estimates," is the most profound of all, but its inherent simplicity is often missed. Look first at the word "requires." The very fact that we place the safety at the end of paths *requires* that after we initially collect task durations to properly size the project, we no longer consider them – *at all.* We are moving away from managing projects at the task level and moving to managing them at the project level. After we have moved the right amount of safety to the end of the paths, we never have to measure, worry, or talk about task durations again. Yes, we will be interested in the causes for *delay* in tasks, but we will never, ever *measure, reward, or castigate* anyone based on task durations again. Yes, we want tasks to be completed as quickly as possible, but we are employing optimal resource assignment, full kitting, active task management (or ATM, see Chapter 21), and delay analysis to accomplish this. We will talk more about this under tactics.

4) To get excellent control, it behooves keeping the number of tasks in the PERT network to less than 300 (for huge projects zooming might be needed).

Why do we want to keep the number of nodes on the PERT to less than 300? *To get excellent control.* Control of what? Control of the management of the project for on-time delivery. As you move beyond 300 nodes (or tasks) for the top-level Project Manager, he or she increasingly loses *focus,* gets distracted, and is pressured to multi-task. Goldratt first learned this concept from Statoil[45], who had a tremendous amount of

45 See the Statoil story in the Introduction.

experience managing megaprojects. It was this experience that led them to "boiling down" a 40,000-task oil rig project to less than 300 tasks.

To some, this might sound ridiculous. How do you *practically* reduce a 40,000 task project to less than 300 tasks? Actually, there is a rational methodology that uses a number of techniques. First, we need to consider how we should model tasks in the CCPM sense of the word. This is discussed in detail in the next chapter. But after we have modeled our project network based on the definitions found there, we also see that "...for huge projects zooming might be needed."

What does this mean? "Zooming," also known as "task nesting" or "sub-projects," means that in order to keep the PERT for the top-level Project Manager at under 300 tasks, we might need to display a series of tasks as a single task on the PERT. Diving down one level (or zooming) a person *separate from the top-level Project Manager* will manage this series or sub-project, and report only the remaining duration for the entire series or sub-project back daily to the top-level Project Manager. From the top-level Project Manager's perspective, this series of tasks or sub-project is only a single task, and the person managing it (perhaps another Project Manager) is simply a task manager (see Chapters 20 and 21). This allows the top-level Project Manager to be focused, free from task switching, and *in control* of the project.

In one of the many conversations I had with Dr. Goldratt about the Projects S&T tree, I asked him specifically about "zooming" (task nesting or sub-projects). He told me that each "zoomed" task is managed just as is any other project, but by a separate person (not the top level project manager) and with one exception. He said that the maximum number of tasks for any sub-project should be 200 tasks – instead of the 300 number for the top level PERT. Theoretically, for a super-massive project, you could even use the zooming idea in sub-projects (in such a case I suppose the maximum number of tasks would be even smaller).

5) Using templates (when applicable) significantly reduces the planning time and reduces unneeded variations.

It is not required that every time we introduce a new project that we build a PERT from scratch. Many projects are repetitive. Some are 100% identical to previous projects. For such situations, saving PERTs as templates is a wise, time-saving course of action. Other projects are not 100% identical to previous projects, but they are close enough that we can begin with a template and modify it to fit these projects without too much time or trouble, moving quickly and saving lead time.

Tactics

For all projects proper PERT networks are built (using templates where appropriate). The time estimates are cut in half and projects and feeding buffers are inserted according to Critical Chain Project Management (CCPM). The projects are properly staggered. Proper actions are taken to ensure that resources are aware that their estimates are regarded as just estimates - they will no longer be judged according to meeting their time estimates. The resulting plan is used to properly release projects into operations. The resulting planning ability is used to determine reliable and acceptable due-date commitments for new projects.

Let's consider these tactics one at a time. There are six of them. They are each powerful and foundational to Critical Chain Project Management:

1) For all projects proper PERT networks are built (using templates where appropriate).

Many organizations do not use PERTs or have detailed project networks. Although these would be an improvement over "back of the envelope" planning, when we come into an organization, we do not try to bring them up to the state of the art in traditional project management before we move on to CCPM. Therefore we do not immediately start building PERTs. Instead, we focus on what is most important - reducing task switching and WIP, making sure resources are optimally assigned, and that we are relentlessly pursuing full kit for active, frozen, and to-be-released projects. During this period we are still executing projects at the same level of planning the company is accustomed to.

However, as soon as these activities are initiated and active, we begin to build PERTs in the correct way (see the next chapter). If an existing project is very close to completion, or if we are sure we will finish the project before we begin executing CCPM, we probably don't need to build a PERT. Just complete the project as quickly as possible and use your energy to build PERTs for other projects. But if there is work expected to be done after launching execution of CCPM, a PERT should be built, at least for the portion of work remaining. Build PERTs in priority order – open projects first, followed by frozen projects, and then to-be-released projects (these projects should still be addressed in priority order). The expectation is that once CCPM is launched, all projects will have PERTs. All PERTs should be saved for future reference. Some should be officially designated as templates.

2) The time estimates are cut in half and projects and feeding buffers are inserted according to Critical Chain Project Management (CCPM).

Please note that nowhere in the S&T tree does it talk about obtaining "aggressive but possible," "50/50," or "challenging" task durations. To do this sends a message we don't want to send. Asking for these types of task durations, although we may say otherwise, still implies that we have an interest in individual tasks being completed "on time," or that tasks have an *expected* duration. Therefore, even though we try to explain that completing tasks on time is not important, we will get resistance ("You're trying to cut my task time in half!") We will also, unfortunately, still get some managers who want tasks to complete at the new "aggressive" pace – and will measure their employees thusly and even penalize them when they are "late." We don't want this to happen! There is no need in having this type of discussion. It is counter-productive.

Dr. Goldratt was adamant about this. Any time he heard anyone talking about "challenging" task estimates or the like, or imposing complex algorithms to calculate buffer sizes, he would angrily exclaim, "Just cut it in half! It's good enough!" I'm ashamed to say it now, but it took me a long time to realize that *Goldratt was right*. While I thought this statement meant Dr. Goldratt was naïve about the "real world," the actual truth was I had missed the Inherent Simplicity in the solution. I was looking for a complex solution to a complex problem.

From the perspective of workers or managers, we are NOT cutting individual task durations in half. *We don't care* what task durations are anymore. We are moving away from managing projects at the task level.

There is one and *only one reason* we ask for task durations at all – to initially and properly size the whole project and its buffers.

Therefore, ask for task durations within the context of what the workers understand – the way they usually commit durations if they commit them at all – with full embedded safety included. Use these durations on the PERT. Let the CCPM software you have chosen later cut them in half to determine the overall un-buffered length of the Critical and Feeding chains and install half the remaining safety as buffers. What could be simpler?

Likewise, there is also usually no reason to have discussions around individual tasks that are process-intensive. "What if my task is a 24-hour cure in an oven? If I say 24 hours, and you cut it in half, I'll be late. If you take half of 24 hours and put it into the buffer, you're making the buffer (and therefore the project) too big." Then we might try to be clever and give the task two-thirds of actual processing time, a 16-hour duration, knowing that the other 8 hours will be added back into the buffer, but the buffer and the project will not be too big. Can you see how this conveys a message that finishing individual tasks on time is still important?

Nonsense. Unless your project is highly process-intensive, doing things like this are trying to manage "within the noise." Doing it the S&T way is "good enough." If your project is one of those rare cases which is really significantly process intensive, why not just cut the percentage of time used for the buffers instead? It's a lot easier and just as effective.

3) The projects are properly staggered.
This tactic is covered in detail in Chapter 18.

4) Proper actions are taken to ensure that resources are aware that their estimates are regarded as just estimates - they will no longer be judged according to meeting their time estimates.

This communication is important even when we are asking for traditional, safety-embedded durations. We don't want people to have the idea that we are going to "cut" their available time and therefore further inflate their already inflated estimates. We must make it absolutely clear that we are moving away from task-level durations – we will not ever hold them accountable for full-safety embedded estimates.

In cases where people have already been "infected" by reading other materials on Critical Chain where the inherent simplicity has been missed, we might have to work extra-hard in explaining how people will not be held to their estimates. If we hear questions or comments due to these sources, we should address them immediately.

5) The resulting plan is used to properly release projects into operations.

The plan referred to here is the properly constructed project portfolio plan – each project has properly constructed PERTs, Critical Chains have been identified and buffers have been inserted, and all projects have been properly staggered off the virtual drum. When used with good CCPM software, the software "knows" when to properly release each leg of each project into work.

6) The resulting planning ability is used to determine reliable and acceptable due-date commitments for new projects.

The ultimate objective of this system is to be a reliable projects company, to deliver on time or early, on or below budget, with full initial intended scope intact. This tactic is the pinnacle of our effort, being able to quote highly-reliable delivery dates for our projects from the very beginning. This means we will also be changing the way our sales people quote lead times and delivery dates. We will, in the future, quote delivery dates based on where we can insert our projects into the stagger of the portfolio (pipeline). This is discussed further in Chapter 18.

Sufficiency Assumptions

Planning is useless unless it significantly helps operations.

And therefore we need further information on building PERTs, transforming the PERTs into good Critical Chain plans, and staggering projects. To do this, we need another level of detail. On to Level 5.

Chapter 16 – 5.113.1 Building Good Project Plans/PERTS

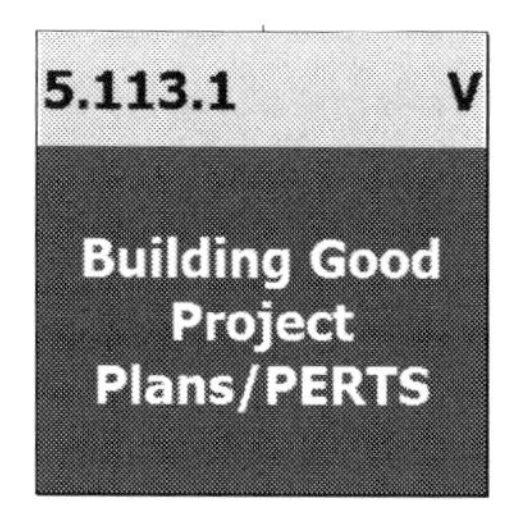

I live in the Seattle area in the United States. We have a professional football team that won the Super Bowl in 2014: the Seattle Seahawks. The team's quarterback, Russell Wilson, who is famous for his meticulous study of Seahawks' opponents, often repeats a saying describing the difference between the winners and the losers, the average and the great, the 'wanna-be's' and the true champions:

"The separation is in the preparation."

In the Japan they have a phrase that conveys a similar message. **Dandori hachibu** means "Planning is 80% of success."

If our goal is rock-solid PERTs for all projects, it would behoove us to have a strong set of strategy and tactics for achieving it. A solid project network yields many benefits – deeper understanding of the project by the entire team, an opportunity to spot significant sources of risk and uncertainty and to plan for them, a mechanism for streamlining the project and minimizing lead time, greater accuracy in determining buffer penetration, task prioritization across projects, and more.

Necessary Assumptions

Managing a project without formal planning (PERT) is a recipe for increased improvisation and miscommunication. Vast experience shows that a PERT that is too detailed (over 300 tasks) is useless as a tool for execution (that is the main reason for neglecting the PERT much before the projects are finished). In most multi-project environments PERTs do not exist or they are much too detailed.

We have already dealt with the concept of keeping the top-level PERT network to below 300 tasks. We have observed a recurring pattern where those who plan projects and those who execute them, appear to be locked in a permanent conflict. In many project environments, PERTs with far too many tasks yield the sort of confusion that causes the people who do the work to abandon the plan early-on and instead, turn to "tribal wisdom" and trial and error techniques in an attempt to succeed. In many such cases, planners complain that "everything would be fine if they (the workers) would only just followed the plan!" At the same time, the workers fire back: "If you knew how things work in the real world, your plans wouldn't be worthless!" Delivery performance usually suffers as the conflict goes unresolved.

Other signs of poor planning in organizations are either the use of makeshift plans or checklists with no logical dependencies, or "winging it" with little or no planning at all. When asked why so little attention is paid to proper planning, the explanation often comes down to a trade-off being made between the perceived effort required first to plan, and then maintain the plan during subsequent execution, and the perceived value to be derived in execution from good planning. High risk and/or complexity would argue for better planning, but the effort required to plan also increases with risk and complexity. So even then, planning often loses out because the perceived effort of planning seems to be too burdensome.

But planning, when done properly, helps to identify and solve potential problems and execution errors before they have a chance to manifest within the project. So how do we tip the scales to the side of good planning? If the amount of investment and effort required for planning and plan maintenance increases in direct proportion to the number of tasks it takes to represent or model the work associated with a project, it is necessary to limit the number of tasks required to manage the project to the smallest number of tasks necessary.

For this we need a set of strategy and tactics – both for the specifications of what a good plan looks like, as well as instruction on the planning process itself – including *who* should be doing the planning. How do we know what is sufficient? How do we know the proper level of detail and task sequence which will help us to execute our projects as fast as possible and avoid errors and rework? Let's find out.

Strategy

All projects about to be released have PROPERLY detailed PERTs.

The strategy, of course, is the "what," not the "how" concerning properly planning project networks. The "how" can be found under the tactics. The benefits of having all projects properly planned should be self-evident, but what we really need to know is how plans can be *usable* by the people who have to do the work. Can we permanently resolve the conflict between those who plan projects and those who execute them? With Critical Chain, we believe the answer is an emphatic, "yes."

The Parallel Assumptions, in this case, provide us with a wealth of information on the "TOC definition" of a task, and the desired level of detail for the network. They are so important that I stress them repeatedly during an implementation, and encourage people to commit them to memory. Building a project network based on these assumptions has proven again and again to open up the lines of communication and create a harmonious work environment between planners and production. The input of the workers in a team context and the discussions with planning on how to execute the plan help get everyone on the "same page."

Parallel Assumptions

Very large projects are managed effectively by relatively small PERTs; the PERTs used to build a north sea oil-rig ($4B) and the overhaul of the largest cargo airplane (the C5) each have less than 300 tasks. The following guidelines can help to tame the tendency to over-inflate a PERT: A PERT is not a task manual. A PERT is not a reminder list. A task that takes less than 2% of the project's lead-time must have a very good reason to appear in the PERT. A task represents a group of work. It should not be broken down to several tasks just because it requires different resources for different durations of time. But it should be broken for chosen key- resource-types; a task should be defined so that those types of resources are required for most of the task time. In most multi-project environments many projects are variations of the same generic project. Using templates (PERTs of generic projects) as the base for constructing the PERT of actual projects, reduces drastically the

required time and efforts and eliminates overly detailed tasks that should not appear in the plan.

The definition of a "task" in a CCPM implementation, and in the structure of a PERT diagram may be different from those in another environment. We use this definition and PERT structure to facilitate the success of a very specific objective: Deliver all projects on time or early, on or under budget, and with full initial scope intact. We should not try to use or change the TOC definition of a task, the project network and its level of detail, or Critical Chain itself for any other purpose. Yes, it is tempting to try to modify CCPM to do other things, but to do so means to compromise the effectiveness of the implementation – to do so is likely to lead to a situation where nobody wins completely and everyone gets less than they desire.

It's ironic that people sometimes reject CCPM because they can't use it (or its software) to perform functions outside its intended parameters – functions based on considerations other than flow, even when they often don't even have such capabilities in their present systems!

Critical Chain is for delivering projects on time – not for collecting costs, carrying out detailed resource management, managing vendors, etc.[46] However, Critical Chain can and does:

1) Produce great cost performance without directly providing for a budget baseline or time and cost reporting during execution.
2) Facilitate excellent resource load to capacity management (demand side and supply side) as well as provide resource managers and task managers the essential information they need to make the best assignment decision on a day to day basis.

---

[46] Some CCPM software products actually help with these and other objectives, but the best ones do it in such a way that there is no compromise and delivery performance is not negatively impacted. If the reader has other business systems that adequately satisfy their additional objectives then it is best to maintain the systems separately as opposed to combining them. The financial and other organizational benefits of a successful implementation of CCPM – in other words of reliability – are too important to be compromised by such integration – especially if poorly executed. With CCPM, flow is always the number one consideration.

3) Facilitate good supply chain management practices by providing key inputs to supply chain decision making.

Now let's take a look at some of the rules and guidelines for properly detailing a PERT network. Under tactics, we will actually put into action what we have learned. Once again, the following guidelines are extremely important and powerful, and the significance of them should not be underestimated:

1) A PERT is not a task manual.

There is a place for a detailed task manual, traveler or resident plan which details a standard operating procedure for everything that should be done. The PERT network is not that place. Remember, the CCPM PERT has one primary purpose only – to help deliver projects on time. In CCPM, PERT entities should represent the general descriptions of the important tasks in the project, as defined below. Descriptions of these tasks should be clear, short and concise, but complete enough so the new worker understands its meaning, or where to go to learn more.

2) A PERT is not a reminder list.

The logic is the same here as in the previous item. PERTs should not be over-detailed with reminder items like "Check out tools," "Check calibration," and "Call inspector." However, many CCPM software products have a very useful feature – a checklist capability for each task. This is the place to put things like reminders. The checklists do not affect the PERT or the schedule. Look for the software product that meets your needs.

3) A task that takes less than 2% of the project's lead-time must have a very good reason to appear in the PERT.

This is an excellent rule of thumb that helps keep PERTs manageable. For example, if a project's total lead time is 100 days, no task less than 2 days should appear in the PERT without a very good reason. What is a very good reason? The answer is up to your best judgment, but let me share an example. While implementing CCPM on a submarine maintenance project (the submarines were serviced while in the water), we determined we should include any task requiring divers regardless of its duration. Why? Because we wanted to be sure that during the execution of these tasks, when divers were in the water, we were able to prevent cranes from lifting heavy objects that might be accidentally dropped into the water, injuring the divers. In addition, we wanted to be sure no one tested the sonar while the divers were in the water, since if the sonar were switched on, the divers would be killed instantly.

Does it sound like a very good reason to you?

4) A task represents a group of work. It should not be broken down to several tasks just because it requires different resources for different durations of time.

As you are building your PERTs, keep an eye open for where you can combine a related series of activities into a single task. Which tasks logically go together? Do this without regard to resource hand-offs. There is no reason to split a task just because a different resource does a relatively minor amount of work within a task (e.g. mid-task inspection). However the Task Manager (see Chapter 21) should monitor the task and make sure all activities are promptly carried out. It is his or her responsibility to report the remaining duration for the entire task, regardless of the resources required.

5) But it should be broken for chosen key- resource-types; a task should be defined so that those types of resources are required for most of the task time.

Key resources are those who are scarce, or hard to get, and if your opportunity to use them is missed, you might have to wait quite a while before they are available again. For this type of resource, it's appropriate to break down a larger task. Just make sure that for the newly created, scarce-resource task, you define it so that the key resource is required for a majority of the task.

6) In most multi-project environments many projects are variations of the same generic project. Using templates (PERTs of generic projects) as the base for constructing the PERT of actual projects reduces drastically the required time and efforts and eliminates overly detailed tasks that should not appear in the plan.

As will be shown in the tactics, a qualified team of experts is required to build project networks/templates. Since these people also have great value elsewhere, it behooves us to use their time building PERTs effectively. Therefore whenever a template can be used rather than building a network from scratch, it should be done. I recommend you keep hard copies of PERTs for templates as well as electronic copies. This is because it's often easier to modify a template if you can unroll it, put it on the wall, and make your changes.

You do not necessarily need to recall the entire project team for a modification of a template, but each member of the team should review and "buy off" modified templates. It makes good sense to highlight your modifications so the project team can easily identify them.

Tactics

All relevant projects (projects which are not to be soon completed and the projects to be released in the near horizon) are considered in order to determine the generic projects. Proper teams construct the templates per each generic project making sure that the resulting PERT will be PROPERLY detailed. Per each relevant project enough uninterrupted time is devoted by the project- planning-team (the key people that constructed the template and the key project people), to PROPERLY modify the template to fit the specific project.

Now we move to the "how to" of building PERTs. The first thing I'd like you to notice is the number of times the word "properly" appears in the S&T tree regarding PERTs. This is both good advice, and a warning that too many PERTs are improperly built, even by people endeavoring to implement Critical Chain. Properly built PERTs help people better understand the project. But more importantly, properly built PERTs help us to have accurate task prioritization across projects. To optimize the entire system and become a truly reliable organization, properly built PERTs are vital.

We'll begin by addressing *who* should be building PERTs, and then we will address *how* they should be built. Keeping this in mind, I'd like to repeat something I shared in the "How to Use This Book" section to help emphasize the importance of not "skimping" on time, resources, etc. when building PERTs:

The fastest way to do anything is to do it right the first time, no matter how long it takes. Likewise, the cheapest way to do anything is to do it right the first time, no matter how much it costs.

I am still amazed whenever I ask an organization, "Who builds your project plans?" and am told something like, "Larry, the planner builds them." Larry is probably a very bright guy. Larry might have once worked on the factory floor. He is considered to have wide knowledge of the organization's functions and operations. But the fact is no single person has or can have all the knowledge necessary to build rock-solid project plans. Granted, Larry is a stellar employee, but he is not omniscient. The evidence that project plans are not being built by the proper *group* of people is that plans are ignored or abandoned in process because they don't fit what is going on in the "real" world. As good as he is, the little room in which Larry toils diligently is not the real world.

PERT networks should be built by a cross-functional team of experts. Representatives from every function effected by the project should be on the team: Engineering, Planning, Production, Quality Assurance, Material Procurement, and every other affected function should take part. This is what the S&T calls a "proper" team. Only a team such as this can put their minds together to create the most "compact" project network.

Training in network building is one of my favorite parts of implementation in companies. In order to raise expectations and ensure maximum participation, I label such training a "Network Building Event." It really is an event. Depending on the size of the organization and the various product lines, I may have two, three or even four teams working together concurrently for two days. It is amazing watching the deep discussions taking place, with everybody fully involved. Everyone has a great time, talking, laughing, and finding innovative ways to execute projects in the shortest possible time. Common comments after the event are: "We have never communicated like this before!" "I can't believe we haven't always done it this way!" and "Now I can see the improvement we've been leaving on the table!"

Contention between engineering, planning and production melts away as consensus is reached on how to plan a project to be executed in the "real world." Just to see the boost in morale across the board makes the event positively worth it.

There are some preparations to be done before building networks. We need a few materials and a lot of empty wall space: A large roll of plotter paper or similar, at least a meter wide. Tape or tacks to affix the plotter paper to the wall. Lots of blank "sticky notes" or similar. Although some facilitators prefer pens because they allow them to see what is written as they stand behind the group observing their work, I suggest a few dozen pencils with erasers – because you may be moving entities around and drawing in new dependency lines. Finally, get some transparent tape to keep the sticky notes on the completed PERTs in place.[47]

The following steps describe the process for properly building PERTs, or Project Networks.

1) Affix a large sheet of plotter paper to an empty space on the wall. This will be the work surface for building your network.

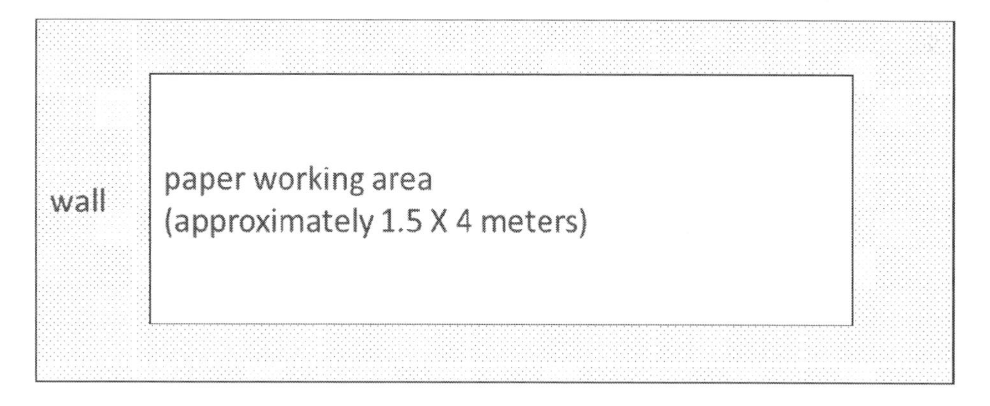

Figure 19. Step 1 - Preparing the work surface

[47] These can be rolled up and stored for future reference without fear that notes will fall off.

We construct a project network that models the *known*. The unknown is absorbed by buffers strategically inserted during Critical Chain Scheduling (next chapter), and MANAGED during execution (chapters 19-24). Please do NOT assume that current practice is best; in fact current practice is often founded on conditions that were once valid but are now no longer sound. Look for opportunities to challenge assumptions ... must we do it this way? Why? Can things be done in parallel? Does the activity in question actually belong on a checklist, rather than on the PERT?

Proper networks are built from the end to the beginning (right to left) using necessity logic. ALWAYS construct the Project Network backwards, beginning at the END of the Project. Why? Because we want to:

a) Capture ONLY the necessary tasks to do the project, and be sure we don't miss any important tasks. We are prone to error when working forward.
b) Ensure correct sequencing and task dependencies.
c) Determine the fastest, most direct route to execute the project

And, should anyone for any reason want to misrepresent a portion of the work, working backwards makes it difficult to do so. Research shows that it is much more difficult to lie if you are instructed to recall events in the reverse sequence to which you claim they occurred. Even though you know the story you are trying to sell, and you are able to sell your lie when telling it in chronological order, your brain will let you down when asked to work backwards from the end. This phenomena comes in handy when looking for creative ways to shorten a project's planned duration. When we plan from start to finish, our brain find it easy to take us down the same old worn path (lie) we have told ourselves. That is, "the way we normally do things is the fastest possible way to do them."

http://www.articlesbase.com/psychology-articles/the-backwards-lying-trap-213765.html

As you construct project networks, standardize the vocabulary and naming convention for parts, equipment, resources, and processes so that these are used consistently throughout the project (and as much as possible throughout the Multi-Project Environment). In other words, if you note a mechanic as "Mech," make sure it's always "Mech," and not

"Mechanic," "Mec" or something else. This will be valuable and help save rework when you enter the network into your CCPM software.

Beginning at the end, you can simply put a sticky note with the words, "End of Project" on it, but I would like to suggest something different – that over the years has proven itself valuable for myself and my colleagues.

The first thing we do is have the group create what is called an "ODSC," which is an acronym for "Objectives, Deliverables, and Success Criteria." It is basically a description of what opportunity is being addressed by the project. It helps to inform the entire team the reasons for doing the project. It also helps to maintain focus. The information for the ODSC can often be obtained from sources such as Project Charters, Project Statements of Work (SOW) and communications / directives from Management.

How does an ODSC help the project team maintain focus? Let's consider a fictional project – say we are building a bridge across a river between two parts of a major city. In a normal project, if you asked project personnel what was the objective of the project, they would probably say, "We're building a bridge." If you asked them to name a deliverable, the answer might very well be, "A bridge." And if you asked about success criteria, they might say, "Success means we'd have a working bridge." But these answers actually do not fit the questions, and show the lack of clarity and focus.

A better answer to the Objective would be something like, "Relief from traffic congestion," or, "A shorter commute." The answer for Deliverables could indeed be "a working bridge," but could also include things like, "Bridge approach roads," "Bicycle path," etc. A better answer to the success criteria question is "City approval," "On time opening," and "Budget met." Here are a few guidelines for each of these categories:

Objectives are what will be achieved by executing this project. The objective should be specifically expressed in Throughput terms for the organization (why are we really doing this?)

Deliverables are those things (usually tangible) produced as a result of doing the project; perhaps a new organizational capability or product, or an element or feature of the new capability or product. Deliverables may be extracted from the scope or SOW.

Success Criteria are the measures by which success is determined for the project, for example: Positive achievements such as due date, cost, or other organizational metrics; boundaries that we cannot overstep, or detrimentally affect, perhaps an existing organizational entity or a condition (e.g. head count can't increase to do the project).

If at all possible qualify and quantify the Deliverables and Success Criteria; the questioning and challenging of these items will bring clarity and consensus; often it brings surprises and new benefits. The discussion itself will be enlightening and valuable for the team.

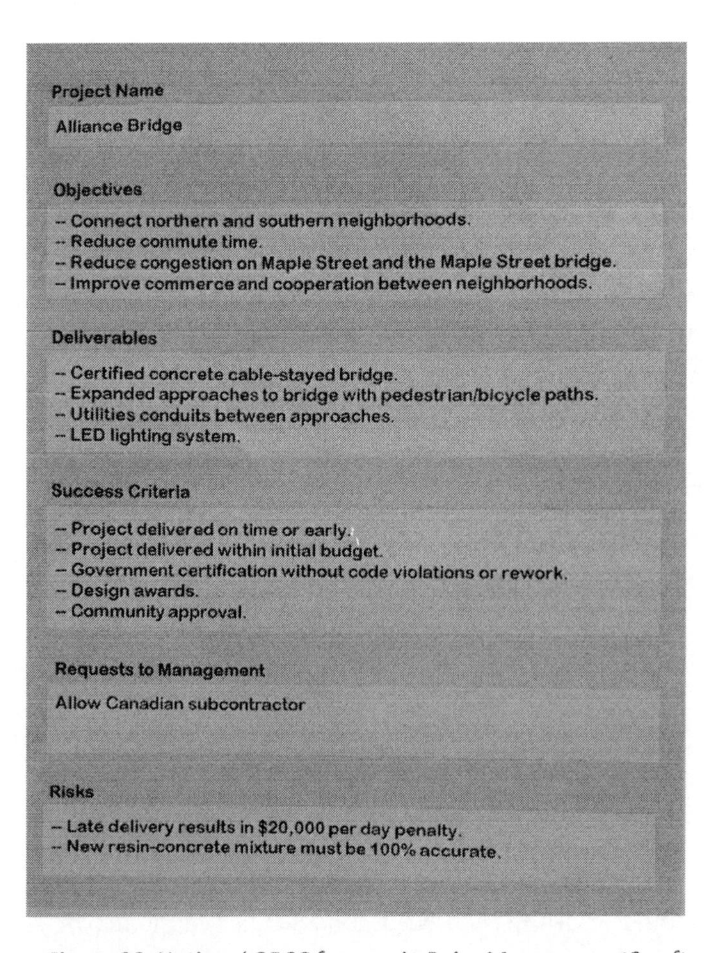
Project Name

Alliance Bridge

Objectives

-- Connect northern and southern neighborhoods.
-- Reduce commute time.
-- Reduce congestion on Maple Street and the Maple Street bridge.
-- Improve commerce and cooperation between neighborhoods.

Deliverables

-- Certified concrete cable-stayed bridge.
-- Expanded approaches to bridge with pedestrian/bicycle paths.
-- Utilities conduits between approaches.
-- LED lighting system.

Success Criteria

-- Project delivered on time or early.
-- Project delivered within initial budget.
-- Government certification without code violations or rework.
-- Design awards.
-- Community approval.

Requests to Management

Allow Canadian subcontractor

Risks

-- Late delivery results in $20,000 per day penalty.
-- New resin-concrete mixture must be 100% accurate.

Figure 20. Notional ODSC form as in BeingManagement3 software.

NOTE: Other information can also be added onto the ODSC form, as in Figure 20 above. BeingManagement3 software adds "Requests to Management," and "Risks." Requests to management could be important if you need something like a new piece of capital equipment. Risks help focus the group as well – they help keep the group vigilant to prevent problems. Taking the time at the start of the project to get people to identify the major elements of risk increases the chance that early detection activities and/or appropriate counter measures are built right into the plan from the start.

Next, assess if achieving the ODSC Statement as finally written will in fact successfully deliver the opportunity. Finally, challenge milestones, if any: Need they exist at all? And if so, need they be fixed dates? Remove them if possible, or convert them to an event NOT tied to a date if that is possible.

2) Using a pencil, on a standard sheet of paper, complete the ODSC information and affix the paper on the far right-hand end of the plotter paper.

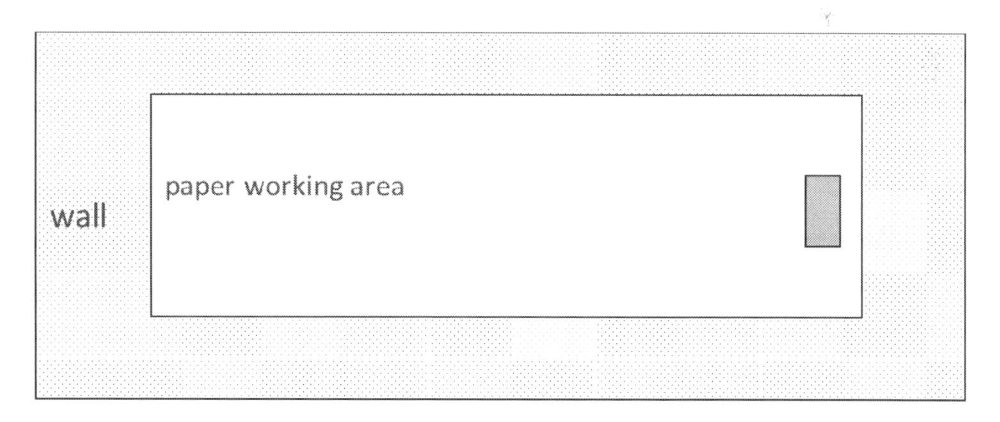

Figure 21. Step 2 – Affix the ODSC statement at the right of work area.

3) Now work backwards from End of Project or the ODSC. Remember we are using necessity logic, meaning we will list the important things that MUST happen in order to complete the project or achieve the ODSC. Use the following verbalization

for this and every entity in the project network: "In order to _________________ (in this case, achieve the ODSC) we must first _________________ .

Maybe this will be something like, "Obtain customer approval." In the case of the bridge, maybe it's, "Ribbon cutting ceremony." Very often multiple things must happen which can be done in parallel before you can start the next task. Make sure all immediate direct predecessors are written on sticky notes. Remember to use a pencil (not a pen!)

You will be making multiple "passes" on the project network in order to make your network as solid as possible. For this pass, only write the task description on the sticky note. In subsequent passes you will be identifying resources and task durations, so leave enough space on the sticky note, or when the time comes, attach another sticky note to the first if you need more room. With each pass, check the network to validate dependencies and look for things that can be done in parallel. We call this "scrubbing" the network.

When one task *must be completed* before we can begin another task, we call it a "task dependency." Validate that task dependencies are truly dependencies – that work cannot be done in parallel. When you have validated dependencies, use a pencil to lightly draw in dependency lines (Figure 23).

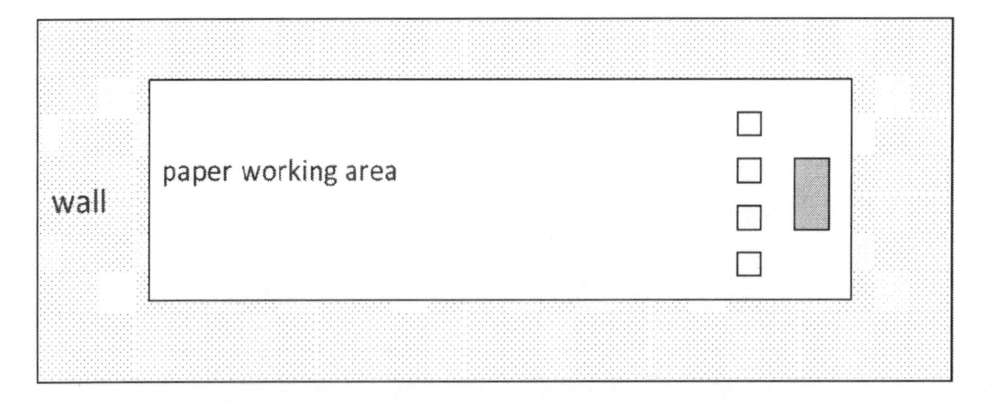

Figure 22. Step 3 - Affix predecessors to the ODSC to the immediate left.

4) One path at a time, for each predecessor, work backwards until the path is complete. With a pencil (not a pen!) lightly draw in the dependency lines. When you are sure you have completed a path, draw a black bar on the left side of the furthest left (1st task in path) sticky note. The black bar indicates there are no predecessors. Continue completing paths, one by one.

5) Each entry task into the network (the first task in a path or chain) is also a full-kit point. Mark all full-kit points with the letters FK and use this as a convention when you transfer to the software. This allows keyword search to identify / filter all full kitting tasks.

At this point, don't worry about the 300 task limit. Just make sure everything important is captured in the network. In the next pass, you can consolidate tasks into logical groups, and/or use task nesting to reduce the top-level network to less than 300 tasks. You can also consolidate tasks if the network has less than 300 tasks, if there are logical groups with no key resources and you don't need to be more detailed.

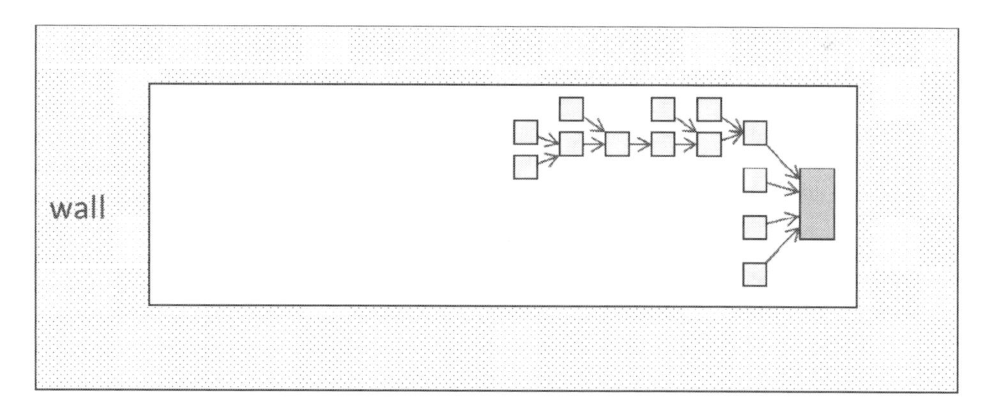

Figure 23. Step 4 –Complete path with dependency lines for first predecessor, then complete the paths for the other predecessors, one by one.

NOTE: If you are including the procurement phase in your network, make sure that the "black box" of the procurement process has been

opened up and the process is detailed with the proper amount of granularity, as was discussed in Chapter 14.

6) When you have finished identifying the tasks for all paths in the project network, you have completed the first pass. Now, make a second pass, but this time start at the far left of the project, and scrutinize the project moving forwards. Is the output of this task needed as an input for any other task? If so, ensure that the link is penciled in. Review the network for risky tasks and procedures; you may wish to move some tasks earlier or later in time.

Count the number of tasks in your network. Are there more than 300? Look for logical groupings of work that you can combine into a single task, even if different resources are involved. Be very sure you have combined as many logical groups of work ("natural" combinations of activities) into single tasks as possible. If you still have more than 300 tasks, consider where you can do task-nesting (also known as zooming, or sub-projects). But task nesting should only be used for huge projects – it is an exceptional (rare) case rather than a common one.

7) Now make a third pass, once again beginning at the end. This time you will identify the resources required for each task. Remember to use standard terminology for resources. Use the resource title, not a person's name. Select the *lowest* skill level that can handle the task. Include as resources all significant equipment, facilities & information (such as cranes, shop floor area, lab equipment, special procedures) not easily obtained, or, for which there is competition (common hand tools would not normally qualify as significant equipment). Make sure that tasks involving key resources are not combined with other tasks. Verify that when a key resource is specified for a task that in fact the resource is needed for the majority of the task; if not split the task into two and reassess.

This is a good point to do a sanity-check. Have the team estimate the overall project duration based on what they know so far. How long should this project take? Individuals may have different answers, so write all the answers on a flip chart or white board. Try to come to a consensus on how long the project should take by asking to explain why they think the project will be shorter or longer. As a result of this discussion, try to settle on a number or a relatively narrow range of durations. Use this as a reference point from which to evaluate the result after adding task durations. If the result is much longer than expected, you should investigate and have a good reason why before accepting the results.

8) After all resources have been identified, make a fourth pass, entering task duration. Allow the workers to quote durations for which they are comfortable. At the same time, remind everyone that no one will ever again be held to task duration estimates. If the subject comes up,[48] tell them you are only collecting durations to properly size the project and the buffers. Explain to them how we are moving completely away from managing durations at the task level, and moving to managing duration at the project level. We are just going to focus on keeping tasks moving once they are started, by making sure they don't start until workers have completed their previous tasks, they have everything they need to start a new task, and that they are not interrupted until the task is complete.

9) Do a final pass before entering the project network into your Critical Chain software. After transfer to the software, gather the group one more time to do a validation check to ensure that the data was accurately transcribed. Then, check once more for task description clarity, task dependency correctness and for omissions by reading backwards in time using "in order to ... we must ..." Check again for resource assignment correctness, and omissions. Challenge assumptions as you go. Finally, make sure you are satisfied that the project network, as constructed, will truly deliver the requirements of the ODSC.
10) Make sure "generic" projects as specified as templates and handled accordingly.

[48] Sometimes people have been reading up on CCPM prior to implementation. If the materials are inaccurate or out of date, they may be worried about their task times being "cut."

The PERT, by itself, will likely be a huge improvement over your current way of planning projects. Now it's time to change the PERT into a Critical Chain project plan.

Was the step correctly implemented?

There is a lot to check in this chapter. It might be best to put yourself into the shoes of the Auditor. What specifically would he or she be looking for? The word "proper" dominates here. First, the Auditor would verify that a proper project team is building networks and templates. This means competent representatives from each functional group connected with the project.

Next, the Auditor would check to see that networks were properly detailed. Was "backwards" planning used in constructing the PERTs? Does any network have more than 300 tasks? Do networks look like they are little more than task manuals? Is there checklist-level minutia in the networks? Does the duration of any task within a network take less than 2% of the overall project lead time?[49] If so, does it have a very good reason?

Is the terminology for resource naming standard and consistent? Have tasks containing key resources been broken down properly? Do task duration estimates retain their original "safety?" Have templates been identified, finalized, and stored in both hard copy and electronic formats?[50]

Were the expected effects realized?

At this point, because we have durations and resources assigned, we also know how long the project is predicted to take from a Critical Path perspective. We can therefore ask questions around how people feel about the validity of the plan at this point. They should have a higher degree of confidence that what the plan is saying is a good representation of how the project is likely to play out.

---

[49] Not generally known with certainty until the Critical Chain steps in the next chapter are complete.
[50] Same as the above.

Using proper project networks has many significant benefits. Most of these are seen in the improvement in the overall performance of the organization, and won't be clearly quantifiable in the short term. So for our purposes here, we should ask this question in the context of more "soft" or less-tangible effects. Is the project team excited about the increased communication? Are they talking about how much better the project networks are now? Is the general harmony amongst the team noticeably better? Are they "sold" now on backwards planning?

What mechanism has been put in place to ensure compliance with the step over time?

Once again, as an organization you should discuss the appropriate mechanisms for each step. Mechanisms should always be centered on the idea of flow being the number one consideration, and therefore should always encourage, rather than impede flow. Your Auditor will likely take a close look at your mechanisms in order to verify this.

The things I would be interested in are having one or more "official" project teams (and the composition of them – including contractors and customers if possible/appropriate) with designated members, the attendance of these members at all network building exercises, the proper assignment and maintenance of templates, sufficient training for new project team members in CCPM and building project networks.

Chapter 17 – 5.113.2 Building Critical Chain Plans

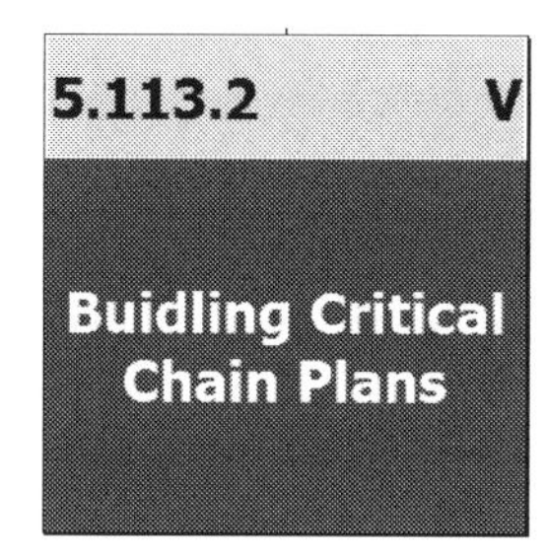

Well-constructed PERTs by themselves are a vast improvement for most projects organizations. But now it's time to turn those PERTs into solid Critical Chain project plans. When you combine multi-project environments and good PERTs with Critical Chain, you have the opportunity to take your organization to a level you might not have thought possible. Almost all projects finishing on time or early, on or under budget, with all the features and functions you hoped for. But to accomplish this, it would seem natural that we also need a solid set of strategy and tactics.

Necessary Assumptions

In most projects the same type of resource is required to perform several tasks. Not considering resource capacity - assuming the same resource can perform multiple tasks in parallel - makes the plan unrealistic to start with and encourages, by design, bad multi-tasking. Having safety embedded in tasks' time-estimates greatly inflates the overall project duration without sufficiently protecting the project completion.

One of the most significant contributions of Dr. Goldratt to the field of project management was to highlight the degree to which the then current state of the art, for all practical purposes, ignored the significant ramifications of resource constraints on determining the most critical path through the schedule. This is not to say resource contention was always ignored. The most savvy project managers developed coping mechanisms to address resource contention. However, these mechanisms created their own headaches by increasing the effort and skill required to maintain schedules of any reasonable size. One positive development since Dr. Goldratt first highlighted the problem is that there have been several studies done and papers published on the topic of how to add this capability to existing scheduling software programs such as Primavera and MS Project. We expect that in the future, all software tools will naturally include this capability. For now, most project managers (and executives as well) remain, even today, largely oblivious to the importance of taking resource contention into consideration during single or even multi-project scheduling.[51]

Also, we have already discussed, as has most available CCPM literature, the fact that at the individual task level, most duration estimates are greatly inflated with safety and that this safety should be removed from within each task and placed instead in more effective locations. Usually, at least 50% of the task duration estimate is protection against variation and Murphy. Due to these two realities, strategy and tactics for dealing with them are in order.

Strategy

The company uses Critical-Chain-PERTs that enable on-time, faster project completion.

The value and benefit of Critical Chain has been proven with vast experience from hundreds of implementations all over the world. The strategy of your project oriented organization should call for CCPM PERTs being utilized for all projects.

---

[51] In addition, after CCPM identifies and resolves resource contention (by moving one or more tasks backward in time), it monitors all chains, including feeding chains, throughout execution, and alerts us when a chain other than the Critical Chain becomes the *most penetrating chain* (the chain currently causing the most Project Buffer penetration), actually usurping the importance of the original Critical Chain (or Critical Path).

Parallel Assumptions

In multi-project environments the same type of resource is required to perform several tasks on many projects. Usually, the specific project know-how gained by a resource that has already worked on the project causes substitution to be inefficient. Therefore, sometimes after the start of a project, even when there is a large pool of identical resources, the number of this type of resources which are practically suitable for a project is limited. Many times a task which is a prerequisite for another task is actually a prerequisite for just a portion of that task. Splitting that task into two components may shorten, significantly, the lead-time of the path. Vast experience shows that the following process provides a realistic PERT for a project: 1. For each key-resource-type the maximum number of resources that will be suitable for the project is defined. 2. The time line of the PERT is adjusted to remove resource contentions. 3. The Critical Chain is identified. 4. On the Critical Chain the possibility of splitting tasks to reduce the lead time is examined. 5. Steps 3 and 4 are done repeatedly until the Critical Chain is finalized. 6. The time estimates are cut in half (not negotiable) and the project buffer and the feeding buffers are created (if there is too much resistance to cut a time estimate in half, don't compromise on the time allotted, instead increase the corresponding buffer)

We have already covered the first part of the Parallel Assumptions in detail, so let's begin in the middle, with a little more information on splitting tasks. In the previous chapter, we talked about splitting tasks to better identify key resources. But there is another good reason to sometimes split tasks, and that is to get shorter projects.[52]

> "Many times a task which is a prerequisite for another task is actually a prerequisite for just a portion of that task. Splitting that

---

[52] As you might have noticed, in many parts of TOC we do not strive for perfection, because at some point continued analysis and adjustments end up only trying to "optimize within the noise" of projects. This is a waste of time and energy, so you will often hear Goldratt and others use the phrase "good enough." However when it comes to project lead time we never want to settle for "good enough." Due to the immense benefits of fitting in extra projects in the same timeframe with the same resources, you should always try to make the Critical Chain networks as short as possible, even when earlier iterations meet delivery date requirements.

task into two components may shorten, significantly, the lead-time of the path."

Once we have identified the Critical Chain and inserted buffers, we might find that our project is still longer than we had hoped it would be. One place to look for shortening projects is in predecessor-successor relationships – i.e. between pairs of tasks on the longest path (Critical Chain) of the project.

Is a predecessor required for the entire subsequent task, or only for a portion of it? If only a portion, this means that the predecessor task is not required to execute the first part of the task – only the second part. If you can find such cases in your project network, you can split the successor task in two and work the first part in parallel with the predecessor and the second part after the predecessor is finally complete. This shortens the project.[53] Even if one of the remaining portions ends up being less than 2% of the overall project duration, this might be a good enough reason to include it.

To illustrate, look at the figure below. In the first instance (A), we see a portion of a chain containing two three-day tasks. Let's assume this is the Critical Chain. After scrubbing the network, we realize that the first three-day task is really only a predecessor for the last portion of the successor task (B). Therefore by splitting the second task into two parts and working the non-dependent portion in parallel with the dependent portion, we can (C) shorten the project by two days. The net result of this is actually a project that is three days shorter, since removing 2 days from the Critical Chain also removes one day of buffer from the Project Buffer.

[53] Remember that if we shorten the Critical Chain, we will also shorten the Project Buffer, so we win two times.

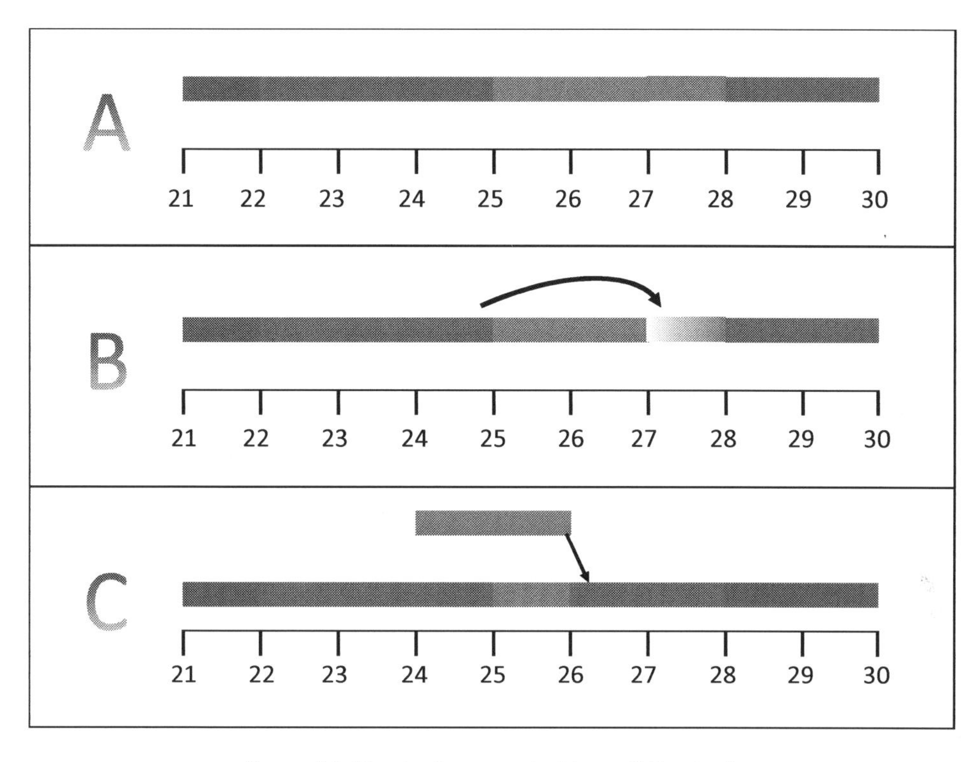

Figure 24. Shortening a project by splitting tasks

Now let's look at the numbered assumptions, which give us further insight on how to turn PERTs into Critical Chain project plans. Please note that for some of these steps (2, 3, and 6), this is the point where software really comes into play. These steps could be done manually, but the software, doing what a software tool does best, helps us do it much more quickly, especially as we are doing multiple iterations of the project network in order to make the project as short in duration as possible.

1) For each key-resource-type the maximum number of resources that will be suitable for the project is defined.

As you enter resources, first on the PERT and then into the CCPM software, this becomes the "official" record of optimal assignment of resources.[54] Double-check once more to be sure you are assigning

[54] It should be understood that there is a difference in CCPM between planning and execution. This is the optimal assignment of resources step in planning. We do this for

resources optimally, not minimally. If you need to remind yourself about optimal assignment of resources, go back to Chapter 8 – 5.111.2, **Accelerate Project Completion** and review.

2) The time line of the PERT is adjusted to remove resource contentions.

Although this activity could be done manually, allowing the CCPM software to do it is much faster and less prone to error. However, because of how the various software products decide which non-Critical Chain task(s) to move back in time, you might want to look at the result and determine if it's acceptable. Some products allow you to make changes if you want the de-confliction to be in a different order.

3) The Critical Chain is identified.

Once again, this is a step that is best done by software. The Critical Chain is the longest path of both task and resource dependencies found in the project network.

4) On the Critical Chain the possibility of splitting tasks to reduce the lead time is examined.

Once you have identified the Critical Chain, you should examine it task by task to see if there is a possibility of splitting tasks in order to shorten the project. This is always a good idea, even if the duration of your project seems satisfactory from a delivery perspective. If you can shorten projects by the maximum possible amount, you might be able to fit in additional projects during the year and reap the rewards of greatly increased throughput.[55]

---

all tasks in the network, on the Critical Chain and on the feeding chains. This is because any chain could become the most penetrating chain (usurping the Critical Chain) in execution. We will address optimal assignment of resources in execution in Chapter 21.

[55] Technically, in a multi-project environment using a Virtual Drum, it is the integration phases that need to be shortened to increase throughput. If the shortening does not affect the integration phase, the number of integration phases that will be completed will remain the same – and therefore throughput will remain the same – regardless of how short or long each project is. But you should still try to find a way to shorten all tasks, since this gives you the most flexibility in determining when to start the project in order to meet a desired due date.

You may only need to do this on the initially-identified Critical Chain. However, sometimes when you shorten the Critical Chain, it ceases to be the Critical Chain because a feeding chain is now longer! As a result you need to re-identify the Critical Chain and go through the process again. Therefore:

5) Steps 3 and 4 are done repeatedly until the Critical Chain is finalized.
6) The time estimates are cut in half (not negotiable) and the project buffer and the feeding buffers are created (if there is too much resistance to cut a time estimate in half, don't compromise on the time allotted, instead increase the corresponding buffer).

This paragraph emphasizes how adamant Dr. Goldratt was about avoiding more complicated approaches such as "Aggressive but possible," ""50/50," or "Challenging" task duration estimates, and employing fancy algorithms to calculate buffer size. In his mind, the quality of the data (which is always less than precise) doesn't justify the increased efforts to be precise in cutting task durations. "Just cut it in half! It's good enough!" was his often-heard advice. Not only are such complications unnecessary, they actually increase people's resistance because sometimes one or more people, in spite of your assurances, will not believe they won't be held to the numbers – numbers the software uses **only** to determine the overall duration of the project. If you are unable to solve this problem, Goldratt suggests you increase the size of the buffers rather than the duration of individual tasks.

Hopefully you won't have to go to this extreme.

Tactics

A CCPM workshop is conducted for all people participating in the project-planning-teams. For each relevant project the project-planning-team continues by following the Critical-Chain process to turn the initial PERT into a Critical-Chain PERT. The templates are finalized.

At the conclusion of the Network Construction Event mentioned in the last chapter, you are ready to enter your first projects into your Critical

Chain software. This means two types of training will be required. The first type is a Critical Chain workshop for the assembled project team(s), called out in the tactic above. This is different from the S&T training that some, but probably not all of the project people have taken. It covers both the theory behind Critical Chain and the specific techniques mentioned in the Parallel Assumptions above.

The second type of training is on how to use the specific CCPM software product your organization has chosen. Some members of the project team(s) might also take this training, as will others not specifically on the PERT-building teams. This training might be conducted by the consultant, or might be conducted by the software company, depending on the situation. In the "How To Use This Book" section, it was mentioned that a "training map" (schedule) be prepared up front for the entire organization, specifying what type of training, who is required, etc. Based on the training needs of your organization, you will determine the best timing of these sessions. But it's preferable that the CCPM workshop is held as soon as the Network Construction Event ends, and that someone already knows how to enter the PERT information into the CCPM software.

The tactic then says that templates should be finalized. To do so there are a couple of more things we need to discuss. These are additional reasons we may need to continue to "scrub" the project network.

First, when the data is entered into the software, the Critical Chain is identified and buffers are inserted, you will know the expected duration of your projects.[56] It's very possible that with removing safety from tasks and adding back half of the sum of safety as buffers, your projects may have become shorter than you thought they were previously. However, especially due to resource de-confliction, you might sometimes find that project duration is longer than you think you can tolerate. Under the current conditions, this is reality, and you can't wish it away. But there are things you can do.

In this case you should roll up your sleeves and make another pass at the network, looking even more closely at things like the possibility of working tasks in parallel and splitting tasks to shorten lead time. You may

---

[56] This does not identify the projects' completion dates yet, just the expected duration. You won't know the expected completion dates until after you implement the staggering step in the next chapter.

also consider things like using loan employees, subcontractors, etc. to reduce the resource contention and make the project shorter.

Second, something that sometimes happens during buffer insertion is that "gaps" can appear in the Critical Chain. The reason is that when feeding buffers are inserted, they push feeding paths backwards in time. If there is a resource dependency between the feeding chain and Critical Chain, a gap can appear. In order to allow for the full use of the feeding buffers, at least in planning, the Critical Chain "waits" for a time. In a perfect world, we would hope for no gaps, because they mean the project has become a bit longer (on the Critical Chain only, no additional time is added to the buffers).

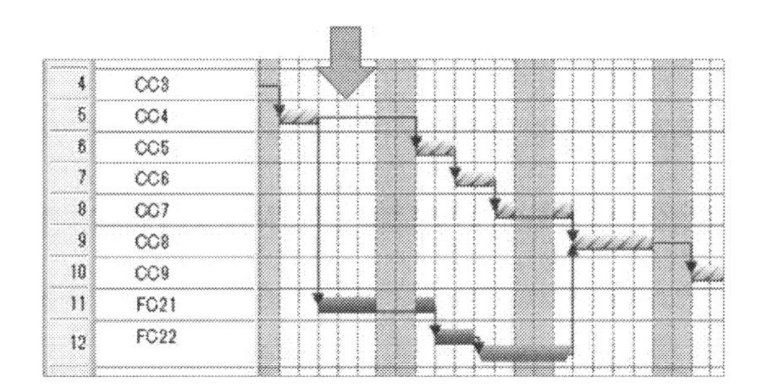

Figure 25. Gap in the Critical Chain caused by resource dependency on feeding chain. In this instance, the Critical Chain is identified by the striped tasks, while the solid color tasks are a feeding chain leading to a feeding buffer.

A secondary effect of this is that a feeding chain can sometimes be pushed backwards in time by a feeding buffer – to the point where the feeding chain may actually be scheduled to start before the Critical Chain. You should make every effort to make corrections to these situations. The techniques to do it are the same as in the above. However, in neither case is it the end of the world if you can't. Just do your best, then finalize the templates.

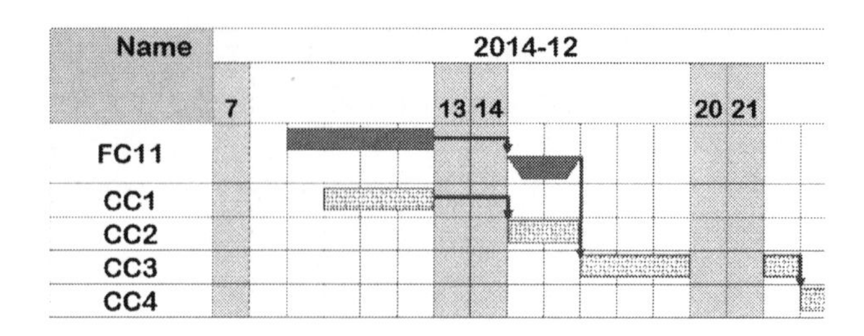

Figure 26. Feeding buffer causes feeding chain to start before Critical Chain

An experienced Critical Chain consultant understands all of this, hence the value he or she can bring to the organization. I've always felt that if you want to learn how to fly an airplane, it's better to have a trained professional sitting next to you than giving it a go yourself after reading a few books. That professional helps keep you from making a mistake you might (not) live to regret.

Was the step correctly implemented?

Once again, it is best to put yourself in the shoes of the Auditor. As you study this chapter, what do you think he or she will be looking for? How about making sure optimal assignment of resources in planning has been properly done for all projects? That all networks have been thoroughly "scrubbed" and are as short as possible – any opportunity for splitting tasks has been exploited. Check for gaps in the Critical Chain and feeding paths beginning before the Critical Chain. Has everything been done to remedy this situation? Verify that templates have been finalized and stored properly, both in electronic and hard copy formats.

Finally, verify the proper training has occurred for the right people – both a Critical Chain workshop and software training.

Were the expected effects realized?

What do we expect from implementing Critical Chain project plans? There should be a consensus amongst your project team(s), and in the larger organization in general, that your project planning has never been better. There should be confidence that lead times, at least at this point, are as short as possible, but at the same time they are completely achievable. Finally, there should be confidence that the plan is executable, meaning the arguments between planning and operations – those who create the plans and those who have to execute them – have disappeared, and that all plans are actually followed as written.

What mechanism has been put in place to ensure compliance with the step over time?

A "buy-off" mechanism involving the entire project team should be established. You might want to measure the instances of any individual refusing to buy-off until changes are made. You may also wish to track these objections and analyze them for patterns or trends, for the purpose of corrective action and continuous improvement. You also might want to keep lists of who has successfully completed the proper training earmarked for each individual, and make sure that additional opportunities for training are available for those who have missed it.

Chapter 18 – 5.113.3 Staggering Project Portfolio

As with all steps in the S&T tree, it is expected that all previous steps will have been successfully completed (using the Three Questions), before proceeding to the next step. Therefore, before staggering the project portfolio(s) using Critical Chain software, 1) Freeze, Accelerate and Full Kit must exist, and 2) Proper network diagrams (PERTs) and proper Critical Chain planning has been done for ALL projects.

In the early years of multi-project Critical Chain, it was assumed that staggering on a heavily-loaded resource would be sufficient to smooth key resource contention. We called using such a resource as a pacing device a resource-based drum. But time and experience have revealed that staggering on a particular resource, although helpful, does not always completely solve the problem of key resource contention. Too many projects are still delayed, and due date performance is often not as good as expected or hoped. The reason for this was discovered to be that in the integration phase of a project, i.e. the phase where the "legs" of the project are merging with the critical chain, there tends to be a frequently occurring concentration of problems (for example rework or engineering changes) that cannot be modeled in a project network.

Therefore it was determined that a better, more consistently effective way to smooth key resource contention had to be developed. The challenge was to achieve the objective of smoothing out the load, without being overly dependent on the accuracy of the underlying resource modeling data.

Necessary Assumptions

In multi-project environments most key resources work across projects. Not considering resource contentions across projects makes the plan unrealistic to start with and encourages, by design, bad multi-tasking.

Why is multi-tasking so common in the multi-project environment? Certainly there are many contributing factors. However, one of the biggest culprits is the way most projects are committed and launched – unconsciously ignoring the resulting load on the *system* (and its individual resources) for a given near-term period. Multi-tasking therefore is a result of bad planning. Resources are unwittingly set up to multi-task due to the failure of the commitment process to protect the resources (by not overloading them). So what can we do about it?

Strategy

Projects are planned (and committed)[57] to ensure effective operation.

This is a rather broad statement, which actually sums up the entirety of what we're trying to do in implementing Critical Chain. But in this step it has two specific meanings: First, it means that our project planning does not create bottlenecks and delays in our operations. We avoid this best by properly staggering projects as a permanent part of our procedures, as described in the Parallel Assumptions and Tactics below. The second meaning (or objective) is that we are able to make reliable due date commitments, meaning we deliver on-time to our initially quoted date, on-budget, and with full scope in excess of 95% of the time. Therefore we also need a new way to make delivery date commitments. We have specific tactics for doing this. Leading us to our tactics are the following parallel assumptions:

Parallel Assumptions

[57] I have added the word "committed" in parenthesis for the purpose of clarity.

An effective way to deal with resource contention across projects is not to try and resolve each resource contention (a futile, exhausting exercise bearing in mind that the actual time the work is performed is likely to be shifted due to the high variability) but rather to do good enough smoothing of the load on each resource type. The temporary peak loads that remain in the plan (and the many more peak loads caused by Murphy) are absorbed by the buffers. A VIRTUAL DRUM staggers the projects in accordance with the actual pace of the system. Therefore, it effectively smoothes the load on each resources type. Emulating the VIRTUAL DRUM in the planning stage resolves the resource contention problem. Emulating VIRTUAL DRUM in planning - the STAGGERING mechanism: 1. For all projects consider ONLY the tasks performed by the chosen integration area. 2. Following the projects priority, place these tasks on a time line, obeying the restriction of number of projects allowed to be worked on in that integration area - Staggering. 3. Adjust the time estimations of the tasks on the time line to reflect the actual rate at which projects finish this integration. 4. For each project use the time determined for the integration tasks as an anchor to place all other activities. 5. Examine the resulting load on key resource types. If there are peak loads that cannot be absorbed within half of the corresponding buffers check for and correct errors in the data. 6. If a certain project is planned to be completed significantly after its committed due-date, better inform the client now.

Although it makes sense in planning to de-conflict resource contentions within a project, it does not make sense (in planning) to attempt to de-conflict resource contentions across projects. Unlike within a single project, where it was important to prevent the overloading of key resources presumed to be dedicated to the project, the approach to managing the load on key resources across the portfolio is approach in a slightly different. In a multi-project environment we can afford to accept a reasonable amount of temporary overload at the organizational or portfolio level during the planning and commitment process.

The reason for this is that no matter how we plan for it, in execution in the multi-project environment reality will be different. What looked like resource contentions in planning will change (due to the various rates of buffer consumption in projects), sometimes dramatically, and when you actually reach that point in time you may find there are no contentions at all. Similarly, even when we see no resource contentions in planning, in execution (again, due to the various rates of buffer consumption,) when we reach a given point in the future such contentions might exist.

So rather than actively try to resolve resource contention across projects through the software algorithm, we will more passively (and effectively) do it through the use of a virtual drum. This is what is meant by "good enough smoothing." "Good enough" so that staggering itself will smooth the load on key resources and the buffers can absorb any remaining peaks of resource contention.

The creation of a virtual drum to stagger projects is designed to help us more effectively smooth the load on our resources, promoting flow by allowing resources to focus and work without interruption. This in turn will contribute to enhancing due date performance. Our first meeting with the virtual drum was in 5.111.3 Defrost, where we created a simple virtual drum to stagger projects until we could get the Critical Chain Steps in place. Now it's time to further define and refine the virtual drum, make it explicit in our PLANNING (and our software), and use it as the mechanism for staggering projects.[58] Based on the stagger we will later be able to determine reliable due dates for all future commitments.

How do we do this? The Parallel Assumptions provide us with a lot of insight:

"Emulating the VIRTUAL DRUM in the planning stage resolves the resource contention problem. Emulating VIRTUAL DRUM in planning - the STAGGERING mechanism." The dictionary definition of "emulate" includes the following: "To attempt to equal or surpass, especially by imitation." Since 5.111.3 Defrost we have a form of Virtual Drum in place. This Virtual Drum has set a new pace for us completing projects. We now have some history of the frequency of project completions. So the best place to start the official, software-assisted stagger is to emulate, or "imitate" the Virtual Drum (the pace of the system) we already have in place.

---

[58] As a validation step for proper selection of the Virtual Drum, make sure that most of the projects in your portfolio pass through all or part of this phase. However if your products go down different lines, through different plants, etc., (i.e. you have multiple mostly-segregated groups of projects) more than one Virtual Drum can be used.

It is easiest to do this in your chosen Critical Chain software. However, some software products do not allow you to identify and stagger on a Virtual Drum. If you have this type of software, you (and your consultant) will have to be creative to get it to do what you want. Let's look at the numbered assumptions for the details on how to emulate our existing Virtual Drum. The tactics below talk about a "proper team" of people with the charter to perform the following and emulate it in the software:

1) For all projects consider ONLY the tasks performed by the chosen integration area.

Now that we have Critical Chain PERTs, we can easily choose and identify the integration area for each project. But it's important we accurately define the term "chosen integration area" in relation to Critical Chain planning.[59] Many organizations routinely use the word "integration" for things such as final assembly. This could include an entire factory floor with work durations of many weeks or even months. It is crucial to understand that this is not the meaning of "integration" in TOC and Critical Chain. Although the "chosen integration area" could indeed exist *within* the traditional integration (or final assembly) area, it does not necessarily refer to all of it.

The broadest definition of integration in a TOC project management sense is found in 5.111.3's (Defrost) parallel assumptions: Integration is the ***phase*** where, for each project, the various legs are coming together. But we can define it further. What is it about the various legs "coming together" that we consider important in defining the virtual drum? What is happening as the legs come together?

First, compounding delays across projects create peaks in demand for key resources, support functions and management attention.[60] To

---

[59] Sometimes the terminology can be confusing. "Integration area" can be taken by some to imply a physical location, which it is not, so we have to be careful. In addition, sometimes we use the term "integration point," which could imply a single resource or task. This also is not necessarily the case. I prefer the term "integration phase," because multiple tasks and resources are often involved in integration. Finally, when the integration phase is used to stagger projects in the multi-project pipeline, it is referred to as the Virtual Drum.

[60] It bears pointing out that there is a consistently large *unplanned* draw on key resources as well – caused by everything from calling key resources to evaluate

operate most effectively as an organization these peaks need to be smoothed out. Next, we find that delays of one or more legs can cause "de-synchronization" within the project, sometimes causing compounding, or cascading delays. It is impossible to plan for this "de-synchronization" in the project network. Also, as legs come together, projects are more prone to problems in fit, form or function. Again, this cannot be modeled in a PERT, because a PERT models the known, and things like de-synchronization and problems with fit, form or function are unknowns. When these things occur, management attention is demanded. Since in TOC we believe management attention (bandwidth) is a primary constraint of all organizations, it makes the most sense to stagger projects based on these *peaks in demand for management attention*. By doing so, the load on key resources is naturally alleviated, because the level of load on the system will be lower by staggering on the phase of increased management attention than by staggering on a heavily loaded resource.

In other words, heavily-loaded resources will be less heavily-loaded. But does this mean we have to choose the entire phase in the project where legs are coming together? Not necessarily. We should have good intuition to know which sub-phase within this larger phase really demands most of our scarce management attention. Where is most of the rework occurring? Where do we tend to appoint "special teams" to attack problems? Where is engineering being called upon to answer questions and resolve issues most often? If we can identify this sub-phase, we have an excellent candidate for our virtual drum. ***This sub-phase occurs as a series of tasks and involves multiple resources.***

KEY UNDERSTANDING: THE "CHOSEN INTEGRATION AREA" (VIRTUAL DRUM) IS THE SERIES OF TASKS WITHIN THE LARGER PHASE OF "LEGS COMING TOGETHER" WHERE MANAGEMENT ATTENTION IS MOST DEMANDED. It can be different for each project. This series of tasks rarely includes the entire set of tasks from where the first leg enters the Critical Chain until the last leg enters the Critical Chain. It should not be a phase which takes months to accomplish.

---

problems on open projects to attending a litany of meetings on every conceivable subject. This is a big factor why drumming on a heavily-loaded resource alone can be misleading, under-stating the true resource load picture –which in reality includes unplanned events.

2) Following the projects priority, place these tasks on a time line, obeying the restriction of number of projects allowed to be worked on in that integration area - Staggering.

This step arranges all projects in the proper order for execution under the new system. List all projects by priority. Next, we will arrange them according to the number of projects allowed to be worked at any given time within the integration area. Remember we decided how many projects should be concurrently allowed in integration during the Defrost step. This was a "best guess" at the time. Review and confirm this number or adjust it accordingly, and continue to list all projects in priority order in groups. Are one or more projects already past the integration area? No problem, just list them first by delivery date and work on them as usual.

Some current projects will be in the integration area. List these in priority order in the first group after the projects which have already completed integration. Once you have listed all current and upcoming planned projects, move on to step 3.

3) Adjust the time estimations of the tasks on the time line to reflect the actual rate at which projects finish this integration.
4) For each project use the time determined for the integration tasks as an anchor to place all other activities.

Now you can begin to build out your stagger into the future. You have the data concerning your actual rate of integration area completions since Defrost. Using this data, determine the date that the next project currently in integration is expected to complete the integration phase, and insert the next project not yet through integration behind it (later on the timeline), ending the phase where the next completion is expected to occur. Since you know the rate of actual integration completions, you can set integration completion points for every remaining current and to-be-released integration phase, in priority order, of course.

As Israeli consultant Dani Omri has pointed out, the pre-CCPM steps (freeze, optimal resource allocation, and full kitting), enable an increased rate of integration phase completion – we do MORE projects at a faster rate compared to the traffic jam we managed before. By emulating the VD in planning, we expect continued improvement and an eventual eclipse of the planned rate. At that point we will make a Virtual Drum adjustment (see Chapter 24).

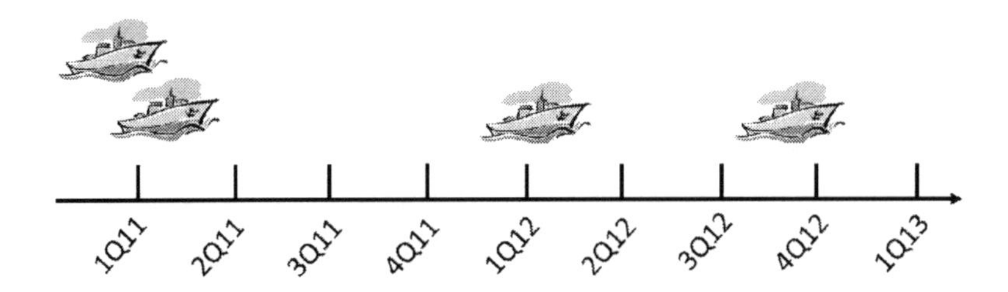

Figure 27. Deliveries before CCPM implementation. Deliveries are not regular - they are usually late, there are few deliveries overall, and they are generally far apart in time

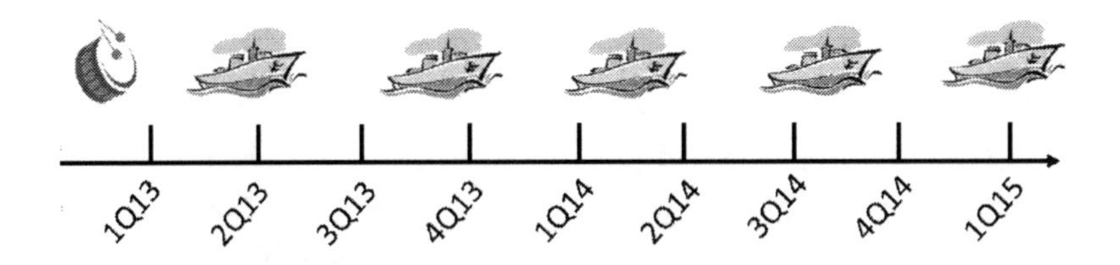

Figure 28. Deliveries after freeze step, with simple Virtual Drum implemented. Deliveries are more regular (approximately one Virtual Drum completion every 4.5 months), closer to on time, with more deliveries overall.

Figure 29. Emulating the Virtual Drum in planning. Using the actual pace of the system (one delivery approximately every 4.5 months), planned projects are laid out on a future timeline. The star indicates the emulation point and activation of the Critical Chain software. Planned project deliveries are in gray boxes.

IMPORTANT NOTE: A major reason we want to emulate the existing Virtual Drum and start with the current actual rate of completion is we already know this is a reasonable, achievable pace. We know this because we're doing it! Therefore we do not want to be any more aggressive than this at first. Once we set the stagger and quote delivery dates to customers, we do not want to ever have to change these commitments to later in time. Therefore we can't afford to buckle under to pressure to stagger and commit based on a more optimistic pace. The good news is that if we do actually perform better than our estimates, and we certainly expect to as time goes by, we can deliver projects early, adjust the Virtual Drum to a faster pace (see Chapter 24) and reap the rewards. Rescheduling the project portfolio later in time (because we were initially over-optimistic) is a nightmare for our reputation that we want to avoid at all costs. Calls and apologies will have to be made to all customers. Understand that re-shuffling projects or rescheduling to a slower pace should be a very rare event, due to the fact that the ramifications are so great. This is why only top management can make decisions that affect the stagger, such as inserting a new, important project in the middle of the pipeline. And in such cases another project *must* be postponed in order not to impact the entire pipeline. If your business reality is such that dropping new projects in at the last minute is unavoidable or even desirable, and you do not have the ability ramp up or ramp down capacity at a moment's notice, then it is crucial that your pipeline planning process make adequate allowances for the random arrival of new projects by reserving a sufficient amount of capacity for such occurrences.

One thing the software will do is insert a capacity buffer. Since the tasks in the integration area have had their safety removed, and we know that variability and Murphy will ensure that additional time will be needed for at least some of the integration tasks, we place a "capacity buffer" between the integration phases for each "lane" in the allowable number of projects in integration. For example, if four projects are allowed concurrently in integration, there are four "lanes." Continuing with the idea that half the safety time removed from a project in integration should be sufficient to protect that project, we usually set the capacity buffer at 50% of the total integration time.

NOTE: It may be necessary to adjust the percentage for the project buffer or capacity buffer upwards or downwards to more precisely match the current pace of the system. It is probable that perfection cannot be reached, just get as close to the current pace as you can.

In other words, since the completion of each integration phase is staggered, the start of each integration phase is known, as is the start and finish of the entire project. This can be seen in figure 30 below.

5) Examine the resulting load on key resource types. If there are peak loads that cannot be absorbed within half of the corresponding buffers check for and correct errors in the data.

Most Critical Chain software will give you resource load histograms similar to those shown in the illustration. Since you are emulating the actuals from your existing Virtual Drum, you have established that your organization is able to maintain this pace, regardless of what the load-to-capacity charts indicate. Therefore, you should not observe in the histograms peak loads on resources that are greater than can be absorbed by 50% of the available buffers. If you do, it is almost always the case that there is some discrepant data influencing the histograms. Therefore you need to find and correct the crucial data errors.

I will address step 6 of the Parallel Assumptions below. Now let's consider the official tactics to use to create and stagger upon a virtual drum.

Tactics

A proper team invests the time needed to emulate the VIRTUAL DRUM and to identify and correct the crucial data errors. Actions are taken to ensure that projects are released according to the plan (legs having different lead-times are released at correspondingly different dates). Actions are taken to ensure that due dates for new projects are committed ONLY according to the STAGGERING mechanism (or top management's decision to postpone a specific existing project).

Let's take the tactics one at a time.

First tactic: "A proper team invests the time needed to emulate the VIRTUAL DRUM and to identify and correct the crucial data errors."

Proper. There's that word again. We need to be certain that a team comprised of people knowledgeable about the entire operation is put together to emulate the Virtual Drum. Only such knowledgeable people will be able to identify and correct errors in the data. Once these errors have been corrected, you should no longer experience peak resource loads that cannot be absorbed by 50% of the available buffers.

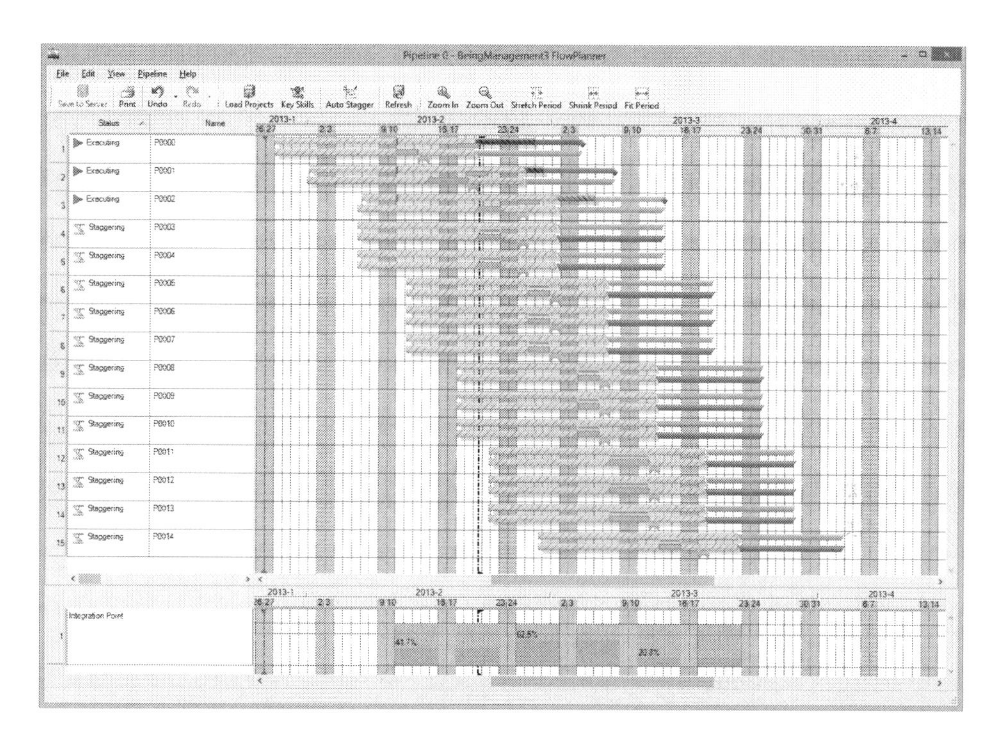

Figure 30. In this view from BeingManagement3 software, projects are staggered with three projects allowed in integration. The top bar in each active project indicates actuals. The bottom bar is the baseline. Orange identifies the Virtual Drum phase. The cyan symbols under each project indicate the capacity buffer.

Second tactic: "Actions are taken to ensure that projects are released according to the plan (legs having different lead-times are released at correspondingly different dates)."

When you have finalized your schedule, the software will identify the start dates, but properly executing this is a team effort involving the Full Kit Manager, resource managers, and task managers. Each of these functions should help ensure that legs of projects are released only according to their start dates.

Third tactic: "Actions are taken to ensure that due dates for new projects are committed ONLY according to the STAGGERING mechanism (or top management's decision to postpone a specific existing project)."

The last tactic here is a crucial one, and represents yet another huge paradigm shift. A big question that people ask when they consider staggering projects is, "How do I quote delivery dates to my customers?" Before we answer this, let's consider that after we have completed the stagger in our new system, we may find that some previously quoted delivery dates for existing projects under contract might be unrealistic, or worse, impossible to meet. Let's return for a moment to our Parallel Assumptions:

6) If a certain project is planned to be completed significantly after its committed due-date, better inform the client now.

In spite of the dates we have committed to the customer in the past, once all steps to this point are properly implemented, we should have, for the first time, a reliable view of our ability to deliver on time. What we are seeing for the first time, is *reality.* Based on our staggering, we may find that some delivery dates are likely to occur significantly later than we originally promised. If this is the case, and we realize that we have erred in quoting delivery dates to our customers, it is best to inform our customers now, while also describing our new process and why we believe we'll be much more reliable with our quotes henceforth. Once again, communication is the key, and most customers prefer honesty and reliability to falsely optimistic estimates that we (or our competitors) rarely meet.

If the new date(s) are completely unacceptable to the customer, the answer is definitely NOT to flood the system again with work. Rather, it becomes a decision of top management (please refer to the tactics in 5.111.1) on whether to freeze an existing project in order to more quickly respond to an important customer, or to bite the bullet and stick to the new date. Due to its ramifications, this type of decision can only be made by someone in charge of all projects.

CAUTIONARY NOTE: Unwillingness to face up to and deal courageously with this situation has caused many organizations to reject this new approach to managing their portfolio of projects. However, those who do tend to find that their fears are much worse than the reality. Most customers are well aware of the poor due date performance of their suppliers and have already taken that into consideration when negotiating a due date.

Now let's return to the third tactic: "Actions are taken to ensure that due dates for new projects are committed ONLY according to the STAGGERING mechanism (or top management's decision to postpone a specific existing project)."

It's a new world. No longer will sales people be allowed to quote delivery dates without first consulting the people managing the pipeline. Using the CCPM software, they will conduct a "what if" analysis to determine the actual reliable delivery date, based on inserting a representative project into the pipeline at the next available "slot" in the next available "lane." This date is quoted back to the customer on a "time-limited" basis for acceptance. If the resultant date is unacceptable to the customer, it is escalated to top management. It can only be a top management decision whether or not to postpone an existing project to make a space available for the new one. What must not be done is to accept the customer's request without postponing another project. Are you serious about being reliable? Then you cannot overload your system ever again. In fact, in a part of the S&T tree not covered by this book, the following tactic is explicit:

3.1.4 Load Control: "**The staggering mechanism of CCPM is strictly obeyed even if it results in losing some bids in the short term.**"

This obviously implies buy-in at the very top of the company, which is why the S&T tree gives us a holistic approach to improvement, and is intended to be taught to executives first.

Here is a quick review of the elements making up the emulation of the Virtual Drum:

1) "Safety included" task duration estimates are made for all tasks in a project.
2) Estimates are cut in half, removing the safety. The Critical Chain is identified and buffers (50% of the safety removed) are inserted. Key resources are level-loaded.
3) The integration phase (where the legs of the project come together) is identified.
4) If possible, a narrower phase, where increased management attention is commonly required, is identified.
5) This phase is identified as the Virtual Drum phase.
6) A determination is made on how many projects should be allowed in the Virtual Drum phase at any point in time.
7) Using the above information, projects are laid out on a time line, with Virtual Drum completions matching the rate at which the system is currently completing projects.
8) Projects are staggered accordingly in the Critical Chain software. If necessary to match the actual pace, adjustments to the project buffer and capacity buffer percentage are made.

Was the step correctly implemented?

I am going to break from the norm here because I want to emphasize the importance of having a complete audit conducted by "outside eyes" at this point of the implementation. As we are now prepared to go into execution, it behooves us to make very sure we have done everything up to this point correctly, and have not overlooked anything or made any errors in implementation that will limit the effectiveness of the solution. A tremendous amount of money, reputation, and internal harmony is at stake, and we should not cut corners or costs at this point. If we do it right, we can make ourselves into ever-flourishing organizations.

Were the expected effects realized?

The expected results of this step are that we have a fully-functional, ready for prime time implementation as verified by the Auditor, without errors in implementation, and an initial stagger based on actual performance. Up to this point we should have seen steady improvement in performance, and the morale of the organization should be significantly better. If any of these things do not exist, we need to re-examine our assumptions, and make corrective actions as necessary prior to moving into execution.

What mechanism has been put in place to ensure compliance with the step over time?

The mechanism for staggering itself is (or should be) incorporated into your Critical Chain software. For new CCPM users, or users who have tried resource-based drumming and are not happy with the results, I encourage you to select a software product that can identify and stagger off of a Virtual Drum. A resource-based drum, although effective, will not yield the same level of results as a Virtual Drum.

Some mechanisms we might consider are a measurement of how many times we are asked to violate the stagger, how many times we actually violate the stagger, how often sales people make commitments without consulting the Pipeline Manager, Full Kit Manager, etc.

Section Five: Day to Day Operations

Flow is the number one consideration

Chapter 19 – 4.11.4 Managing Execution

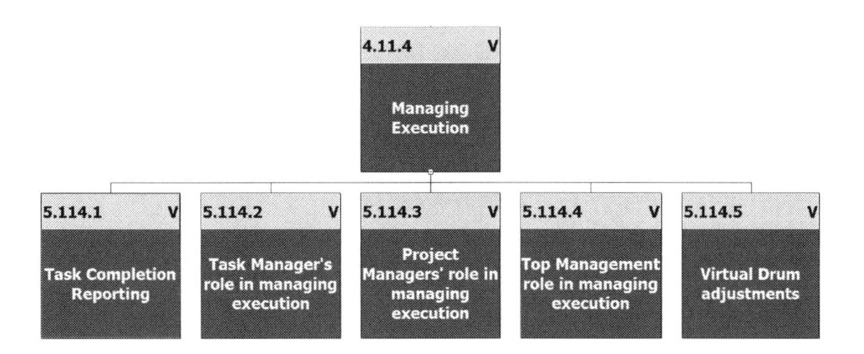

At last! We are finally ready to reap the rewards of our weeks of hard work. Although we have been seeing steady progress since the Freeze step, our improvement is akin to gaining speed on a highway ramp. We are now ready to enter full-speed on the highway to ever- flourishing. Just as we needed strategy and tactics in our planning, we also need solid strategy and tactics for execution.

Necessary Assumptions

Hectic priorities result in a "crisis mode" of management. The combination of the above two phenomena delays needed management assistance.

Crisis management is so common for most of us that we make jokes about it. Much of the chaos we experience today is actually caused by the way we manage – reacting to client pressure, starting projects as soon as contracts are won, etc. The atmosphere of constant crisis causes us to ignore important things like preparations for the never-ending stream of "urgent" issues. The value of the distinction between important and urgent is that working on important things has the effect of reducing the number of eventual crises we have to deal with.

While working only on the urgent can *never* lead to stability, TOC is are all about focus. It is about reducing the load (and thereby reducing the urgent), and avoiding the unimportant. It is also about focusing management on what our real priorities are. Focus allows management to direct their attention to things that will truly help the system.

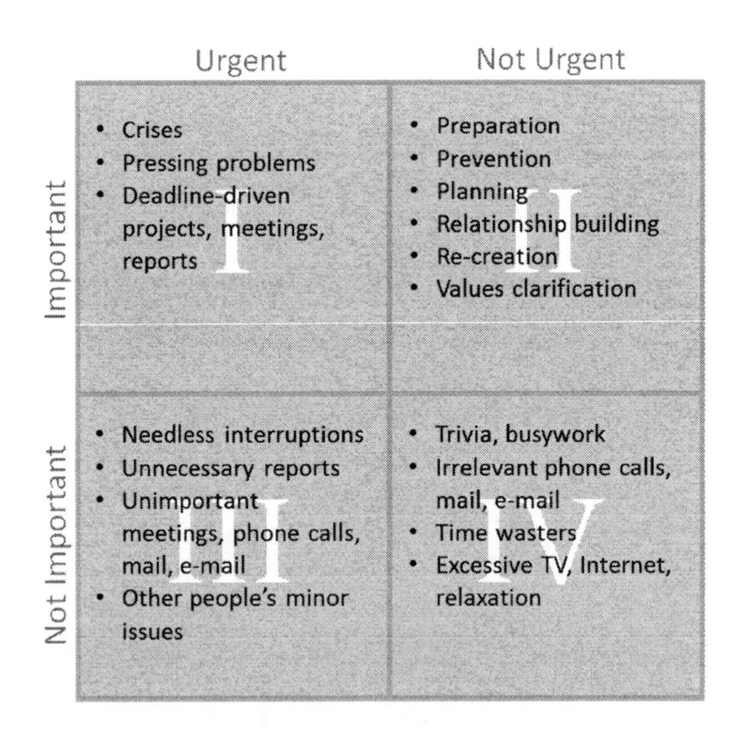

Figure 31. Covey's Time Management Matrix

For many, this may bring to mind Stephen R. Covey's Time Management Matrix (see figure 31). The Matrix consists of four quadrants, which presumably contain all the things which consume our limited time. Crisis management (or fire-fighting) "lives" in the upper-left quadrant of the matrix. TOC and CCPM help us to live in the upper right quadrant, and to be able to more easily discern those things that are neither important nor urgent so we can avoid them. It always makes sense to work on the important things before they become urgent.

Strategy

Projects are actively managed to ensure their successful, rapid completion.

I believe it is safe to say that in the past, our projects have been managed re-actively. The chaos in our systems has allowed us only to react to problems once they surface. Therefore, completing projects has been a combination of reaction and force, pushing our projects to completion. This strategy is about active management, guiding rather than pushing, and acting rather than reacting to bring our projects to rapid completion. This strategy is the "what" of the step. The "how" of the step, is detailed in the tactics at this level and in the steps at Level 5. Parallel Assumptions, describing aspects of our reality, help lead us to our tactics.

Parallel Assumptions

The only way to determine the priority of a task is by examining its impact on the completion of the project. In other words, priorities should be set ONLY according to the degree the task is consuming from its project (or feeding) buffer. Critical Chain Buffer Management is a priority system that operates according to this concept. Management assistance can usually help a top priority task. Helping top priority tasks is helping the projects. The assistance that can (and should) be provided by task manager is different in nature from the assistance that can be provided by project manager. Top manager's assistance is sometime indispensable.

Once again, there is a lot of information packed into the Parallel Assumptions. The first assumption is a bold one, asserting that the only way the priority of a task can be objectively determined is by examining its impact on the completion of the project. All tasks are important – importance of a task does not change regardless of changes in the day-to-day circumstances. However, we all know that no matter how well we plan, nothing will play out exactly as anticipated. This could lead one to assume that the only way for plans to be useful is if we modify our plans every time something changes in reality. The prospect of spending a lot of time updating plans to keep them relevant is daunting or downright scary to most organizations.

We advocate a different perspective and hence technical approach. This approach is to introduce a task prioritization system that does not rely on the need to constantly be maintaining our plans, but rather uses these plans in a new and creative way to provide real-time guidance on where to focus resources on a day to day basis.

Task priorities provide an indication of the relative risk to the due date posed by each task, at this moment. This is a dynamic temperature measure that is updated as conditions ebb and flow on a day to day basis. It therefore needs to have a credible underlying basis for its computation but also needs to be something that can be easily updated without losing its integrity.

In multi-project Critical Chain, task priorities are set across all projects according to the degree that individual tasks impact their project's buffers. The task with the most buffer penetration, regardless of which project it belongs to, is the highest priority. The task with the second-highest degree of buffer penetration, again regardless of the project, gets second priority. Only those tasks that are endangering a project's on-time delivery are subject to special attention. As a result, management of a project portfolio boils down to two simple words: Buffer Management. This is a paradigm shift from the traditional idea of managing projects by scheduling and trying to manage every individual task. By utilizing the impact on buffers to set task priorities, a clear, system-wide, unambiguous list of what is truly needed today can be created. This list equals *focus* for management.

Since project managers can often provide assistance for tasks that task managers cannot, the focus provided by Critical Chain allows project managers to concentrate only on the small number of tasks that need attention (and help) today. And since top management can often provide assistance that project managers cannot, knowing the true priorities allows top management to direct the necessary resources and assistance to the places where they will have the most positive impact, lightening the atmosphere of high demand on their scarce management bandwidth.

Sharing buffer status and the impacting tasks across the organization allows instant escalation of issues to the level in the organization chart where effective action can be taken.

Tactics

Critical Chain Buffer Management is the ONLY system used to provide priorities. Priority reports are provided in different forms to different management functions. Mechanisms are set to enable proper usage of the priority information.

From this point forward, it will be the official policy of the organization that the priorities produced by the Critical Chain system will be the only priorities followed. No more private priority lists by task managers, project managers or production managers. No more priority lists based on phone calls from sales or other departments, or who is yelling the loudest. Since we know we have the most accurate, objective, and reliable prioritization scheme possible, we will not violate this policy.

Most CCPM software products will provide priority lists based on function. Task managers can get lists that only show what is in their area or what is coming to them soon. Project Managers can get lists that cover every task and all task managers in their project. Full Kit Managers, resource managers, and others can get complete lists of all the tasks within a portfolio, etc. In addition, everyone usually has access to view the multi-project task priority list, which shows all tasks in priority order, across all projects.

The mechanisms to be set to enable the proper usage of these lists are detailed in the steps at Level 5. All of these lists provide focus for each function, and the global lists provide an opportunity for those with less urgent (relatively lower time sensitivity) tasks to provide assistance to those with the most urgent (relatively higher time sensitivity) tasks. For the first time, the entire organization can be on the same page, and more than that, can cooperate between departments to benefit the company and everyone involved.

Sufficiency Assumptions

Knowing when not to intervene is almost as important as knowing when to intervene.

Another valuable benefit of Critical Chain is that in addition to the fact that the task prioritization system provides information on what are truly our priorities, by way of Green/Yellow/Red Buffer Management we also know what tasks are not urgent and do not have to cause us concern.

At Level 5, we will see not only how this works in our day-to-day operations, but also how the roles and responsibilities of different groups of people working in synchronization lead us to the objective of all projects being delivered on time, on budget, and with their full original scope intact.

Chapter 20 – 5.114.1 Task Completion Reporting

It is an unfortunate fact that many organizations use and do well with the planning aspects of CCPM, but fail to take advantage of, or simply misapply its powerful execution aspects. In my mind, the execution part is where the "rubber meets the road," and all the CCPM planning amounts to is preparation for successful execution. It is execution that delivers the >95% delivery performance numbers. It is execution that transforms the organization into an ever-flourishing enterprise.

Here, at Level 5, are the rules and guidelines for successful execution of CCPM. As with the planning aspects, everything is designed with flow in mind as the number one consideration. To protect and increase flow, there are a few basic things we must do as a matter of daily practice and policy. It all begins with a foundation of task completion reporting.

Necessary Assumptions

Variability (and its big brother Murphy) changes priorities. In most multi-project environments, frequent reporting on progress by task managers is constantly demanded. Still the frequency and accuracy of the reports is far from satisfactory.

It is a fact of life that things don't always go exactly as we have planned. Nowhere is this more apparent than when executing concurrent multiple projects, where depending on the environment, we could have 5, 50 or 500 projects in work simultaneously. Things go wrong. Machines break down. People are sick, in training, or otherwise unavailable. And when things are not where they are supposed to be at the appointed time, management cannot objectively assess the progress of projects. Therefore they continually ask for more and more detailed status.

Many of us are familiar with hours-long status meetings, consuming the time of every task manager, where literally every task in work is reported upon, and action items are given for those tasks behind schedule. In fact, sometimes it almost seems that three-fourths of a task manager's time is consumed collecting status, and the other quarter is consumed by reporting it. This leaves before and after official work hours for him or her to get their real work done. Worse, a lot of the status demanded is for the purpose of financial accounting, and therefore is a waste of time from the task manager's perspective. And yet with all this reporting, and all these action items, projects still fail to come in on time, on budget, and with their original scope intact.

Obviously, without some kind of status reporting, we couldn't tell how our projects were doing. We would be lost, and everything would seem to be totally out of control. There must be a better and easier way of collecting status, one that will make our operations meetings much shorter and much more focused. One that will allow task managers to ensure tasks are completed correctly – one that will allow Project Managers and executives to know the true status of all projects. To find this way, we need a solid strategy and the tactics to accomplish it.

Strategy

The required data is always adequately available.

What a wonderful strategy! Surely things would be much better if the appropriate data were always available in near real-time. But how do we make that happen? To guide us to the proper tactics, let's look at the Parallel Assumptions.

Parallel Assumptions

People tend to procrastinate on their reporting when reporting doesn't have an immediate/significant impact on them. Traditionally the things

demanded to be reported by task managers are used for financial purposes (calculating the cost absorbed by the projects). In multi-project environments this use has no relevancy to the task managers. In multi-project environments the pressure, exerted from all sides, makes it very important for task managers to know the true priorities. The data that is essential to determine priorities is not the amount of time already invested in a task but the estimation of the time still required for the task to be finished (task status). A delay in a task (and an expected delay) can change the critical chain resulting in a major change in priority to tasks of many task managers. Conclusion: when there is a proper priority system, daily reporting on tasks' expected completion dates is extremely helpful to task managers.

In spite of the constant quest for status, rarely is complete status ever accurately gathered. Workers are frustrated and annoyed by constant requests for status. Numbers on an accountant's spreadsheet are meaningless to them (and to the task managers), and since the choice is between reporting status and performing work, work usually wins. Task managers are often called away to fight fires (and attend status meetings). By the time of the actual status meeting occurs, the progress of tasks may have changed significantly. New problems may have arisen, meaning the completion of the task will be delayed. On the other hand, tasks may complete sooner than expected.

Confounding the situation, without a good priority system action items resulting from rudderless task status reporting result in misappropriated use of everyone's time and create even more task switching. Our exhaustive efforts to collect status actually delay our projects even more.

One of the brilliant concepts revealed by Eli Goldratt in the development of CCPM concerns the nature of status reporting. The important question is not what has been done, i.e. how much progress has been made, but how much remaining work there is to do. This is important because the length of time it took to work on the first part of a task has no bearing on how long it will take to complete the last part of a task. A new problems may arise, or an additional resource may become available. Therefore, the most germane question to ask when collecting status is not,

"What percentage of the task is complete?" but "How many days do you estimate remain until this task is finished?"

In Critical Chain, the answer to this question has direct bearing on buffer consumption, and therefore has a direct effect on the priority of a task. When you supply CCPM's priority system with complete remaining duration information, the result is a dependable task priority list drawn from the best information we presently have. In addition, you now know which tasks are truly urgent, and which tasks are less so, thereby providing focus for any necessary action items.

Tactics

Proper explanation is given to all task managers: what is required from them to report on a daily basis, how this information is going to be used and that they will, at last, be able to obey ONLY the formal priority list. The company launches the daily reporting (by task managers - not by the resources) procedure and relentlessly enforces it.

When task managers are given a full explanation of what information is required, how it is to be used, and what the benefit is for them, they are much more likely to expend the necessary efforts to collect status. The benefits go beyond simply having a single, reliable priority system. Task managers also benefit from the focus the priority list brings, meaning they can direct work on the true high-priority tasks while paying less attention for the moment to tasks that are not endangering the on-time completion of the project, and that their status meetings will be much shorter and effective.

The good news is that it takes less data, effort and time to report on time remaining on a task than it takes to explain, justify or disguise a lack of progress (and to avoid blame). The time saved can then be used to address the issues impeding progress on the most urgent tasks.

It is critical therefore, that remaining duration be reported on a daily basis for all active tasks. Failure to report daily means the data becomes stale – and the buffers, and therefore the priority list, become increasingly unreliable. Therefore the S&T tree once again uses the powerful word, "relentlessly," as in relentlessly enforcing daily remaining duration reporting. One other thing to note is that workers should be allowed to *work*, not report status, so this responsibility falls on the task manager. There is also an important role played by the forced daily interaction between TM and worker that is typically lost when worker reports their own status.

Since task managers will be getting priority lists from the CCPM software and will need to make remaining task duration inputs in the software, in addition to the training given to task managers based on the content of this chapter and the next, appropriate software training must also be delivered.

Was the step correctly implemented?

Taking "blame and shame" out of the organization is key to CCPM implementation and the first step in that direction is reflected in implementing this tactic. So I want to emphasize once again, if this step is not implemented correctly, stop. Do not continue to the next step. You will want to check whether task managers have been trained properly both in reporting daily remaining task duration and in your chosen CCPM software. But the main way to tell that this step was correctly implemented is to verify that remaining duration for all active tasks is being reported by all task managers on a daily basis. A suggestion on how to do this is described in Question 3 below.

NOTE: Please be aware that some people can potentially use the reporting system, once they understand it, in their favor by creating false remaining duration "emergencies" in the system in order to "push" their project. This behavior must be highlighted and prevented. A "false alarm" can be easily exposed by proper investigation in Buffer Management meetings.

Were the expected effects realized?

The expected effects of this step include the daily production of a dependable priority list that is being utilized as the organization's only guide to priorities. There should also be an overall positive attitude about reporting remaining duration, and an eagerness to do it properly on a daily basis.

What mechanism has been put in place to ensure compliance with the step over time?

The best compliance mechanism is a measurement of the percentage of active tasks being updated on a daily basis. For measurements such as this, I have always favored general organizational measurements over personal measurements that call out violators by name. If the level of daily updating falls below a certain threshold, for example 98%, a general reminder is given to all task managers to be sure to update the remaining durations daily.

This measurement is sometimes called an "Update Index." Your CCPM software might help you with this measurement, or a simple script can be written to extract the data you're your software and load it into Excel to perform the necessary calculations. The Update Index is a high priority since daily remaining task duration updates are crucial for reflecting accurate buffer consumption, thereby providing reliable priority lists.

Chapter 21 – 5.114.2 Task Manager's Role in Managing Execution

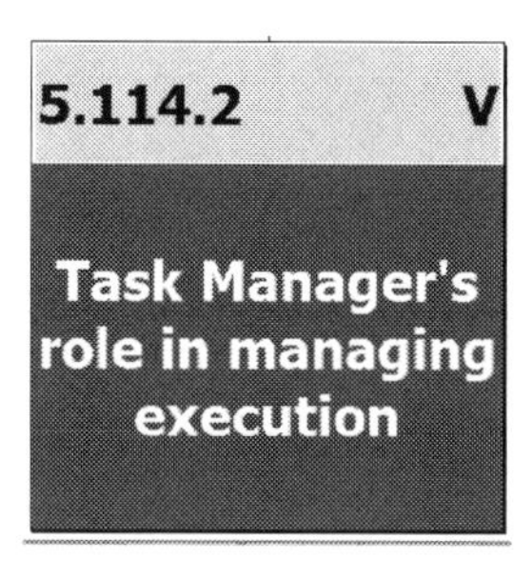

Now we will begin to look a little more specifically at roles and responsibilities for daily execution in a Critical Chain environment. In this chapter we will detail further the task manager's role in managing execution. For those unfamiliar with the term, a task manager is someone who has responsibility for assigning the optimal number of workers to tasks, collects and reports daily remaining duration of tasks, works to ensure task-level full kit, and follows up to make sure the tasks have been completed successfully. In some organizations they may be called leads or supervisors. Task managers do not necessarily do the work themselves. They are responsible for an area or specific part of the work, and for the purposes of project management would report to the applicable project manager(s).

We have already covered daily updating of remaining task duration, but there is more for the task manager to do according to the S&T tree. Still, Critical Chain is not about adding workload to anyone. Rather, we strive to replace chaos with calm – firefighting with simplicity – and give the task manager more time to do what is important.

Necessary Assumptions

In the common "crisis mode" management, task managers shift priorities frequently and intervene in tasks' execution mainly when it is clear the task will not be completed on time.

Since few if any systems other than Critical Chain have accurate and dependable priority lists, it is understandable that task managers are constantly preoccupied with shuffling tasks and rearranging priorities in order to complete tasks and deliver projects on time. If task managers are to escape from this trap, we need a set of solid strategy and tactics.

Strategy

Tasks are executed according to their priorities. Preparations and corrective actions are taken in due time.

Like most of the strategies in the S&T tree, the "what" of the step is simple and straightforward. Basically, our strategy from now on is to execute priorities only according to the list generated by the CCPM software and received by the task managers. Since we always know which tasks have the highest priority, frequent interruptions of tasks and reprioritizing are not necessary. This frees up an abundant amount of the task manager's time – time which can then be used to do task-level preparations and take necessary corrective actions when problems arise, as well as to take preventative steps to minimize the future fires.

To identify the "how" of this step, we will look to the Parallel Assumptions to help us determine the best tactics.

Parallel Assumptions

Vast experience suggests the following: Based on status reporting, each task has its up-to-date priority according to the impact it has on its project completion - percent penetration into the corresponding buffer. Every day the task manager gets two lists of tasks: The list of tasks currently being executed and the list of tasks that are incoming, both sorted according to their up-to- date priorities. Based on the tasks' priority of currently executed tasks the task managers, aimed at minimizing/eliminating delays, decide on the level and type of intervention. For each incoming task, the task manager ensures the necessary conditions to start the task are in place: approvals, designs, (uninterrupted) resources etc.

When all task managers update tasks regularly, a major benefit is that each task manager is immediately informed when a predecessor task is completed. When all predecessors of a task are complete, a task is considered to be ready to start. Based on the remaining duration input of all task managers the previous day, each task manager should receive two lists generated by the CCPM software. The first list will contain all of the active tasks in a task manager's area of responsibility.[61] This is the list of tasks that can be worked on today, in priority order.

The second list will contain tasks that will be coming into the task manager's area in the near future. This list, also in priority order, give the task manager the opportunity to make task level preparations so the incoming tasks can commence and proceed without interruption.

In summary, a task manager's job is to:

1) Ensure all active tasks in the task manager's area of responsibility are making good progress and are updated daily
2) Ensure all "ready to start" tasks are assigned the optimal number of resources and are started promptly
3) Ensure all upcoming tasks complete preparations (including task-level full kit)

Now let's consider the definition of task priority. Tasks are color-coded as follows:

1) "Black tasks," if any, are the highest priority, since black indicates the project is likely to be late. Task managers should offer any possible help to the worker(s) executing black tasks, and should seek help from his or her superiors if necessary.
2) Next in priority are "red" tasks. Red indicates that there is a danger of a project being late, although we still may be able to recover. For all intents and purposes, they are treated almost exactly like black tasks. Task managers should do whatever it

[61] Please note these are not paper lists, but electronic lists. Even though percentage of buffer penetration is usually only calculated once per day, task completions and new tasks that have moved into a task manager's area will appear in most CCPM software in real time, making the electronic list much more current and valuable than a paper list.

takes to complete these tasks as soon as possible, and hopefully even recover some buffer.

3) Yellow and then green tasks should be worked as time permits. But there is no reason to take any special action for these colored tasks. A yellow buffer indicates the project is healthy, but we should keep an eye on yellow tasks in case they turn red. For these tasks, a task manager needs to be ready to take corrective action if necessary. Green tasks mean there is little buffer consumption, and we don't need to pay any special attention at all to these tasks. This is not to imply that we do not work on green tasks. In a system under control, the majority of tasks will be green because they are progressing smoothly.

NOTE: When we initially stagger projects, we make sure that all the remaining work on open projects have complete, unpenetrated buffers. Therefore, all buffers start out as green. As time goes on task managers should almost always have time in their day to work on yellow and green tasks, after completing any black and red tasks they might have. If a task manager finds him or herself working only on black and red tasks, it is an indication that either we did not freeze enough projects initially, or we have a Department that has a unique overload situation (see Chapter 29 for what to do if the latter is true.)

Before we move to the tactics let's spend a moment talking about task-level preparations (or task-level full kit). Task-level preparations are different from the project-level preparations we talked about in chapters 11-13. They normally involve minor items that are needed to complete a task, such as hand tools, calibrated measurement devices, or supervisor approvals. Although seemingly small and unimportant, lack of adherence to task level preparations can lead to task delays and sometimes delay a project. One example comes from a major industrial company in India. A certain type of inspector was in high demand, his time was scarce and had to be carefully scheduled.

The inspector arrived at the appointed time to review some work done on a large steel product. The work in question was located at the bottom of the product and the inspector had to get underneath it for a clear view. However when he crawled underneath, he found that it was dark and he was unable to see the work clearly. He asked the workers in the area if they had a flashlight, but none did. Being in high demand, the inspector had to move on to his next work assignment, and told the workers in the area he would return when time permitted and a light was available.

The project was delayed for several days before the inspector could finally return. For lack of a minor item (a flashlight), the project was stopped dead in its tracks. A light for the inspector to use is a good example of a task-level preparations. It would not make sense to include a flashlight in project-level preparations. Utilizing a preparations checklist, either in the CCPM software or even on paper, would have saved project from days of delay.

As the task manager reviews the lists of tasks waiting to start or shortly to be arriving in his area, he or she should ensure that all task level preparations have been completed and all tasks can proceed without interruption once started.

Tactics

Following the priorities, task managers assign the optimal number of resources to tasks. Task managers review daily two lists of tasks (open and incoming) and according to the up-to- date priorities make sure tasks are effectively progressing.

The first tactic talks about task managers assigning optimal resources to tasks. In Chapter 8 and elsewhere we have talked about optimal assignment of resources in planning. Since variability and Murphy can cause any chain to become the most penetrating chain, we plan for all tasks to be optimally resourced. But in execution we cannot always staff all tasks with optimal resources. People call in sick, are required to attend training or safety sessions, may be on loan from green tasks to help on a red project, or may be otherwise unavailable. One of the responsibilities of the task manager therefore, is to endeavor to ensure optimal assignment of resources in execution.

WARNING!!! People will sometimes ignore buffer-based priorities because the temptation to peanut-butter spread resources across all

available tasks is too great. DO NOT allow this to happen. Some active tasks may have to wait in order to optimally assign resources to the high priority tasks. These are lower-priority tasks, which by definition means they have more buffer protecting them from delays.

Black and red tasks should always have optimal resources assigned. It is less important for yellow and green tasks to have optimal assignment of resources in execution, but we still try to do it whenever we can. But when a task manager is faced with limited resources, it is necessary for him or her to allocate those resources first to the most important tasks. Some organizations even consider optimal assignment of resources part of task-level full kitting. We think this is a great idea and encourage you to consider it.

The second tactic talks about the task manager making sure that tasks are effectively progressing. In addition to daily reporting of remaining duration, optimal assignment of resources, and task level full kit, an excellent tool for the task manager to use is something called Active Task Management (ATM). Although there may be other definitions of active task management, for our purposes it involves the task manager asking each worker three questions for tasks to which they are assigned:

Active Task Management

1) Do you have everything you need to execute this task (task-level full kit)?
2) What is the criteria for knowing that the task is finished?[62]
3) What can I do to help you finish this task faster?

The first question is to verify that task-level full kit is complete and in place. It also gives the worker an opportunity to add something to the task-level full kit. The second question is more important than you might think, because different people often have different ideas on what constitutes a completed task. Does completion only pertain to the work that needs to be done? Or does it include the successful result of an

---

[62] This information is supplied by the task manager to the resource - not the other way around. The task manager should set and confirm expectations for what "done" looks like, and for what should be accomplished daily. ATM also includes passing the baton successfully between task managers, including confirmation of acceptance by the successor.

inspection? Does it include the worker or task manager moving the work to the next Department in the CCPM software? Does it include moving it physically? Completion of a task should be clearly defined and practiced consistently.

The third question is the most important of all. It involves the task manager asking the worker, "What can I do to help you complete this task as quickly as possible?" There are some things that the worker is unable to resolve that the task manager may be able to. Also, some workers may feel that asking for help is a sign of weakness. So rather than waiting for the worker to ask for the task manager's help, it is much better for the task manager to be proactive and offer his or her help up front. With these three simple questions, Active Task Management provides great assistance in bringing in projects on time.

Was the step correctly implemented?

If this step has been correctly implemented, task managers are consistently assigning optimal resources to every possible task, ensuring that at the very minimum Black and red tasks have optimal resources assigned. Task managers are daily reviewing two lists: the list of active tasks currently in the task manager's area, and the list of tasks expected to enter the task manager's area soon. For the first list, task managers are consistently using Active Task Management. For the second list, task managers are ensuring task level full kit, are keeping an eye open for the red and black tasks that might be entering their area, and making sure that the hand-off is smooth.

Of course, this is in addition to ensuring that remaining duration for all active tasks is updated on a daily basis. Another key role of the task manager is to keep task-level work-in-progress low.

Were the expected effects realized?

The main expected effect of this step is that all tasks are able to move without interruption at the maximum possible speed through the task manager's area. Worker morale should also improve since instead of just being told what to do, the task manager is a worker's proactive team member, ensuring full kit (including optimal resource assignment) and offering his or her help as required for each and every task. WIP should be low, giving the task managers and the workers the ability to focus.

What mechanism has been put in place to ensure compliance with the step over time?

Most CCPM software products offer some way to record delay reasons for holdups in task execution. How often are tasks delayed because they do not have task level full kit? How often are successor tasks delayed because a task has not been adequately completed in fact or in the software?

In addition, you might want to monitor the number of times that high-priority tasks which are ready to start are kept waiting while skills are working on lower-priority tasks, and watch for negative trends over time.

Chapter 22 – 5.114.3 Project Manager's Role in Managing Execution

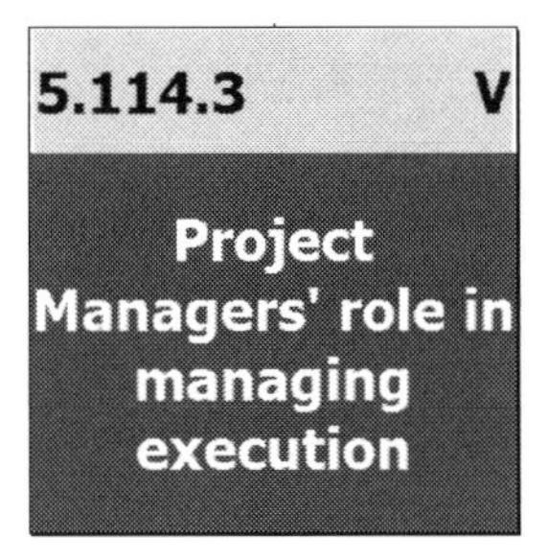

In the last chapter we talked about task managers, the people who are responsible for moving tasks forward, optimally assigning to tasks the people they have been allocated, and reporting daily remaining duration. In this chapter we will talk about project managers, the people ultimately responsible for the financial, quality, and delivery and scope performance of their projects. Typically, organizations employ project managers who have the responsibility and (a measure of) authority to manage their projects to this criteria. However, some organizations execute projects without an official project manager. Others designate people as project managers who really do not have the responsibility and authority to meet these goals – project managers in name only.

Still others assign multiple projects to individual people they call project managers, overwhelming them with too many tasks to monitor and virtually hamstringing them from being able to deliver on their charter. In a Critical Chain system, each project should have an official project manager with the requisite responsibility and authority, and no single person should be responsible for monitoring more than 300 tasks in total.

When the day to day responsibility for task management is shifted from the project manager to the task managers (via ATM), the project manager is freed to do a better job of monitoring and responding to larger, more strategic concerns.

The real paradigm shift in this step is the move away from the project manager attempting to manage a project by managing individual tasks, to managing the project through the management of buffers.

Let's take a look at the logic behind the roles and responsibilities of the project manager in a critical chain implementation.

Necessary Assumptions

There are cases in which task managers cannot take an effective action to minimize/eliminate delays (the required corrective actions are outside the task manager's control or effective influence).

Since task managers are limited in their ability to solve problems relating to their task responsibilities, the project manager is often called on to assert his or her control or influence to keep the project from being delayed. For example, the project manager may be able to speak directly with a subcontractor or supplier and negotiate for a more favorable situation than a task manager could. Therefore due to the project manager's responsibility, authority, control and influence, we need a strategy for guiding this important position during the execution of a project.

Strategy

Project managers are driving a "project buffer recovery" process for cross departmental actions and exceptions not handled by task management.

The sentence is packed full of useful information about the project manager. First, we are told his or her primary responsibility is to "drive a project buffer recovery process." Critical Chain is unique in that, just as buffers can be consumed, they can also be recovered. It is the project manager's job to monitor the project's buffers and intervene as necessary with the tools he or she has available. To do this, the project manager often takes or negotiates cross- departmental actions and agreements which are beyond the authority of the task manager.

Parallel Assumptions

At any given point in time the task that determines the completion of the project is the task that penetrates the most into the project buffer. When adequate reporting is done, an up-to-date report is available that lists, for each project, the tasks that penetrate the most into the project buffer (and also provides visibility into the status of the feeding buffers). Using that list a project manager knows which tasks are essential to check with the corresponding task manager if proper actions have been taken and if help is needed (and therefore also knowing which tasks do not require intervention).

Daily remaining duration reporting, resulting in an up-to-date and accurate view of buffer penetration, is essential to the success of the project manager. The priority list specifies which tasks are currently most penetrating the buffers. The amount of buffer penetration informs the project manager as to whether or not actions, recovery plans, or requests for help from management may be necessary. It is the daily responsibility of the project manager to monitor the buffers and the priority list, and to discuss high-priority tasks with the appropriate task managers. Armed with the knowledge received, the project manager knows where to begin in an effort to recover buffer.

Tasks that are green or yellow need no immediate scrutiny, allowing the project manager to focus on the true priorities. One factor that is proven its usefulness over and over again, is a formal, weekly buffer management meeting chaired by the project manager, and attended by a sufficient number of task managers. This is where the true extent of delays and problems are discussed, and where recovery plans are formulated and followed up.

It is important that time is not wasted in the buffer management meeting by reporting on or discussing the detailed status of various tasks. This is to be done by task managers routinely, outside of any formal meeting. This meeting is about assessing project buffer status and developing recovery plans.

Tactics

Project managers review daily the list of tasks penetrating the most into the project buffer and check if recovery actions are taken or required to ensure that the project is effectively progressing. In extreme cases the project's Critical Chain PERT (and even the template) is updated.

The health of the project is a project manager's primary concern. In the past, in order to determine this, the project manager has asked for greater and greater detail – sometimes literally reviewing the progress of every task every day. This is the antithesis of focus. CCPM can be the best friend of the project manager by allowing him or her to get relevant information at a glance. The project manager gets this by looking at project buffer penetration (based not on original task duration or the percentage of tasks completed, but on remaining duration) and reviewing only black and red tasks. Healthy buffers and tasks do not need review beyond the normal ATM process.

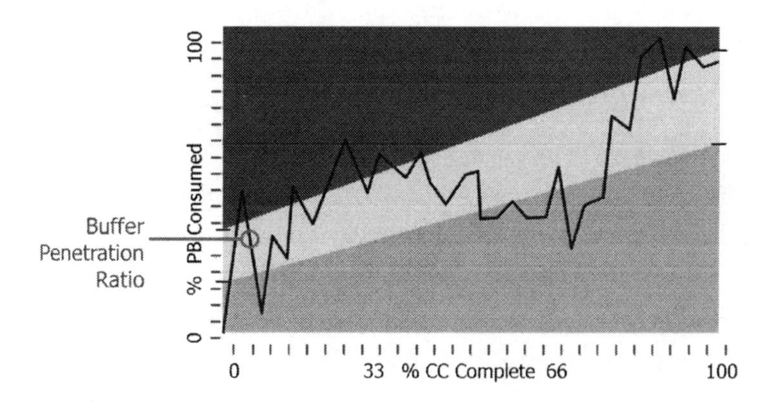

Figure 32. A "fever chart" for a specified project. The black line indicates the history of buffer penetration percentage in the project. It is assumed action was taken each time the red zone (top zone) was penetrated, allowing buffer to be recovered.

The information the project manager receives from buffer penetration helps determine the priority and level of recovery action. At this point the weekly Buffer Management Meeting becomes key. If a project buffer is red, it means less than desirable buffer is remaining relative to the remaining work to be done on the project. This means that the best way to have a reasonable assurance that the project will complete on time is to recover "enough" buffer. As project buffer status is reviewed, the knowledgeable people to create a buffer recovery plan are already in the room or can be invited. By analyzing the most penetrating chain, the amount of buffer desired to be recovered can be determined.[63] The reason you want to look at the entire remaining chain is that it is highly unlikely you will be able to recover sufficient buffer in the current task or even the next couple of tasks. Once you know the desired amount of buffer to recover, the team can put their heads together to determine ways to recover as much of this amount as possible. A number of questions can be asked. Can tasks be split? Can additional resources be acquired? Can overtime be worked? Is outsourcing an option? And many more. As the leader of this effort, the project manager fulfills the intent of the strategy, driving the project buffer recovery process.

Additional things the project manager will be interested in: Are tasks being updated daily? What are the buffer trends? Are optimal resources being assigned to tasks? Are resources being multi-tasked? Is full kitting being followed/completed? Are recovery plans being aggressively executed? And relating to the last tactic – do problems in the project relate to a flaw (or improvement potential) in the project network and/or template? If you are able to create a robust buffer recovery plan, does it make sense to use the process all the time and document it in the template?

Finally, the project manager should borrow a question from Active Task Management. "What can I do to help you finish faster?" Rather than pounding on task managers to work harder, the project manager must be an essential part of the holistic solution. As the strategy indicates, there are things that the project manager can handle that task managers cannot. This is a key to the cultural shift that will help move the company from unreliability to reliability.

---

[63] If you are near the beginning of the project, you will want to recover as much buffer as possible – ideally back to green. If however you are near the end of the project, it may not be necessary to recover the buffer even into yellow, but just enough to feel that barring a major catastrophe, you have "enough" buffer. Remember, if a project finishes in the red, it is still ahead of schedule. Only projects that finish in the black are truly late.

Was the step correctly implemented?

If the step has been implemented correctly, you have official project managers with responsibility and authority to positively influence their projects toward maximum delivery, budget, quality and scope performance. They are responsible for monitoring less than 300 tasks. They hold weekly Buffer Management meetings with task managers and together they devise buffer recovery plans as necessary. Project managers monitor and follow up on these plans.

They have received the S&T, CCPM and software training appropriate for their position. They understand and contribute to the holistic improvement of the organization, and offer help to task managers as well as other project managers (when their project is healthy and another's is in trouble). Buffer recovery action plans are formally recorded and kept current for each project.

Were the expected effects realized?

The expected results of this step have both immediate and longer-term implications. In the short term, it has an impact on task manager knowledge and morale, and creates more of a team approach between him/her and task managers. In the longer term, driving buffer recovery plans means projects are experiencing shorter delays, more on-time deliveries, and better cost/quality/scope performance.

What mechanism has been put in place to ensure compliance with the step over time?

This is again a good place to discuss mechanisms within your organization. You could measure frequency of Buffer Management meetings. It might also make sense to measure buffer recovery and adherence to buffer recovery plans. There should be a formal mechanism to log buffer recovery action plans and to collect summaries for all active projects.

Chapter 23 – 5.114.4 Top Management Role in Managing Execution

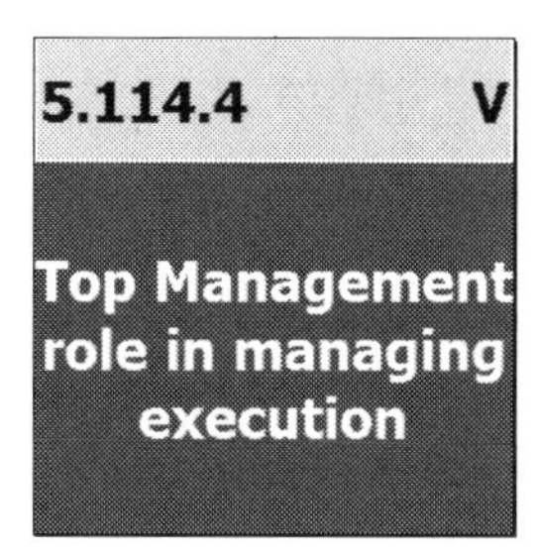

NOTE: In larger organizations there is often one or more additional layers of management between project managers and top management, and some organizations have a Project (or Program) Management Office (PMO). In such cases Buffer Management Meetings should be held at the appropriate intervals with Project and/or Portfolio Managers in attendance, and the content of this step in the S&T generally applies.

Projects are of extreme importance to all organizations, even to those who do not consider themselves to be projects organizations. It could reasonably be argued that the success of an organization is keyed off the successful execution of both internal and external projects, and that the success of a manager is derived from the success of his or her management of projects. IT projects, safety projects, quality projects, HR projects, marketing projects, and many other kinds of projects make up the day-to-day operations of every kind of modern business.

Therefore it is remarkable when you understand that many organizations do not have a cohesive, holistic approach to project management. This is especially surprising for companies that deliver projects as their main form of revenue. The involvement of top management, as well as the alignment of strategy and tactics with every level of the organization separates the best companies from the also-rans.

Necessary Assumptions

In most multi-project environments top managers don't have good enough visibility into the projects. On one hand they are bombarded with requests for more (resources, equipment etc.) and on the other hand projects that seemed to be progressing well are (all of a sudden) reported as going to be late and then very late.

In project organizations, top management is often accused of "being out of touch," changing priorities seemingly at random and causing general chaos as workers have to shift from one project to another to satisfy today's emergency request. Yet when you examine the issue closely, it is easy to see how this situation becomes reality. Since top management, like everyone else – lacks a focusing tool – something that shows them the true health or lack of health in the projects they are responsible for, they do their best to respond to the cacophony of requests coming from all sides. As they have no solid basis from which to act, every phone call from an irate customer generates a chain of phone calls down management lines which eventually ends up changing priorities on the shop floor.

In addition, since projects are often measured based on task progress or utilization of budget, rather than on remaining duration, projects tend to abruptly become late, and then as the assumption says, very late. To avoid this situation, it is obvious a solid set of strategy and tactics is required.

Strategy

Top management is well informed and in full control.

In a mere nine words, this short sentence outlining the strategy contains two profound ideas. The first sounds simple but is rarely attained: "Top management is well informed." The only way this can be accomplished is if top management has a process to extract only relevant information from the data ocean and the myriad of competing voices. The second part of the strategy, that top management is in full control, means that the decisions they make do not negatively impact the overall system. In other words, the ramifications of their actions are known in advance and are designed to minimize damage. As usual, the "what" of the step sounds idealistic, if not absurd. To discover the proper tactics for turning the ideal into reality we look to the Parallel Assumptions.

Parallel Assumptions

In most multi-project environments a project's progress is judged according to the percent of hours already invested relative to the total hours planned. (The result is the prevailing phenomenon where 90% of the project is done in one year and the remaining 10% takes another year.) The effective measure of a project's progress is percent of Critical-Chain completed. The measure of a project progressing well is the percent of Critical-Chain completed relative to the percent of project buffer consumed. The measurement of project recovery is the improvement in the previous measurement. A suitable graph gives a one page, clear picture on all projects current status (and for each specific task that currently determines the project completion date).

The primary role of top management is to remove roadblocks to smooth project execution. They must obtain relevant information and take the proper actions (guided by the principle of flow being the number one consideration). There are a number of assumptions here, all of which are subject to testing and challenge. However, in our experience these assumptions have been validated over and over again. The first part explains the problem of trying to measure the health of projects based man-hours (or man-days) consumed as compared with hours or days planned. Assessing a project based on invested effort to date is like trying to drive a car by looking in the rearview mirror. Since this method tells us only what we have done, and gives us little information about the quantity of work remaining, managers are left to make important decisions based on an incomplete picture of reality. A valid indication of how well a project is going must be based on an assessment of what it will take to finish.

Next comes an assumption that a much more effective measure of a project's progress is the amount of Critical Chain that has been completed, followed by a second assumption that a project's health (in terms of on-time delivery) can best be measured by relating the amount of critical chain completed with the amount of project buffer consumed.

This yields a buffer consumption percentage that can be measured at any point in the project. Relative to one measurement, a lower buffer penetration percentage at a subsequent point indicates improvement or recovery of a project. Likewise, a higher buffer penetration percentage indicates deterioration of project health.

Finally, a graph showing the health of all projects within a project portfolio can be drawn on a single page. Assuming remaining task durations are updated daily, top managers can receive fresh and trustworthy information on all their projects at a glance. This means top managers can know whether they should look immediately more deeply into their subordinates (or customer's) requests, or whether these requests do not require immediate executive action.

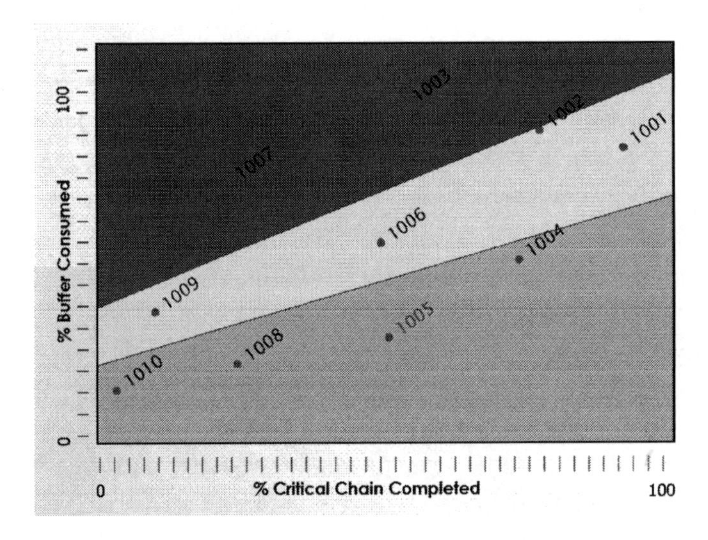

Figure 33. An example of a multi-project fever chart

In addition, buffer penetration percentage allows one to calculate the priorities of tasks across all projects – i.e. creating an unbiased mechanism to show the relative urgency of each active or upcoming task, regardless of which project it belongs to. This means that everyone from executives to workers can always know the highest-priority task within the entire portfolio, the second highest-priority task, and so on. Armed with this truly relevant information, top management can make well-informed decisions that help preserve and enhance the stability of the system.

Tactics

Top management reviews periodically (every two weeks) the projects' status. For projects whose progress is not satisfactory, the recovery actions are examined.

It is recommended that top management (or the PMO) hold Buffer Management meetings no less frequently than every two weeks. These meetings may be attended by project managers, portfolio managers, or both. It may also be beneficial to include selected functional managers both for their knowledge and insight, and for following up on action items that may be created in the meeting. Similar to project manager's buffer management meetings, where only black or red tasks are examined, in this meeting only black or red projects are examined. Since a project manager's job is to drive buffer recovery plans, not only the problem but also the solution should be presented to top management, including multiple alternatives if appropriate.

At this point, top management should make a determination on the adequacy of these plans, and should also borrow the third question from the Active Task Management questions: "What can I do to help you compete your project faster?" Similar questions include, "What is blocking you? What are you waiting for?" And, "From whom do you need help?"

Last but not least, since top management will now have a holistic view of their projects, knowing which projects are healthy, which projects need attention, and which specific tasks are most important to complete, project managers of healthy projects can be directed to offer assistance to the project managers of projects in trouble. When project managers experience the benefit of the help coming from their peers, the willingness and likelihood that the favor will be returned is greatly increased. This is a tangible expression of teamwork and cooperation, creating win-wins for all project managers.

Was the step correctly implemented?

Obviously the correct implementation of this step once again depends upon the timely reporting of remaining duration for all tasks in the system. Like the project managers, top management should pay attention to system's update index, and make every effort to ensure that rules for daily remaining task duration updates are strictly followed. You'll want to verify that bi-weekly executive (or PMO) buffer management meetings are indeed being held, the presentation of up-to-date multi-project fever charts, and the attendance of all necessary project and/or portfolio managers and other necessary personnel. Also verify that buffer recovery action plans have been prepared prior to the meeting and are being presented as a natural part of the discussion, and that action items are routinely and aggressively followed up.

Look for the identification of resource bottlenecks affecting multiple projects, and for frequently reoccurring delay reasons.

Were the expected effects realized?

The expected results of this step match the strategy: Top management is well informed and in full control. Top management no longer makes reactive changes to priorities, and when approached by subordinates for additional resources, etc. can use project buffer penetration and key resource information to make consistently better-informed decisions. Top management is seen as part of the team, aligned with everyone else for the holistic good of the organization, and therefore are no longer accused of being "out of touch." Their decision making is perceived by everyone to be timely and effective. Requests for help are fulfilled. Most importantly, top management is offering and providing assistance when needed and according to priority, thus increasing the flow and performance of the entire organization.

What mechanism has been put in place to ensure compliance with the step over time?

In team discussions you can determine what would be the best mechanisms for your organization. You may wish to consider mechanisms that measure the frequency of Buffer Management meetings, how often they are being postponed, whether they are promptly rescheduled, the presentation of buffer recovery plans and the swift execution of action items. You may also wish to measure employee morale and their perception of top management.

Are executives actually attending these meetings? What are the number of executive decision requested? How many decisions are made, and how many are deferred?

Chapter 24 – 5.114.5 Virtual Drum Adjustments

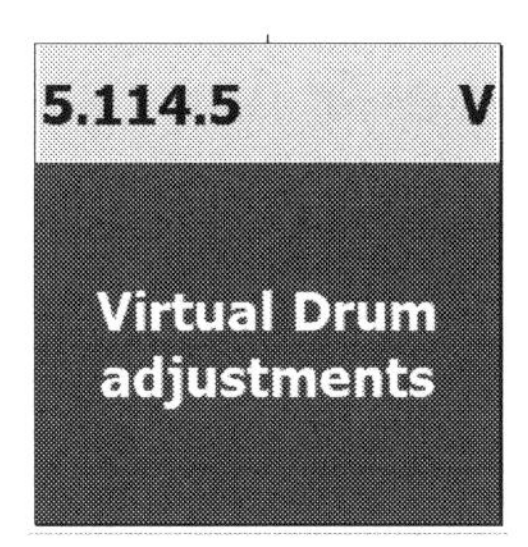

The final Level 5 entity under 4.11.4 Managing Execution concerns adjustments to the Virtual Drum. Ideally, adjustments to the Virtual Drum should be a result of improvement in the performance of the organization. The need for adjustment is the validation we are significantly improving. Without this validation we can't be confident that we are on the path of continuous improvement. It is the mechanism by which we book or bank the improvement beyond the initial burn-down of the backlog.

The need for such adjustments means the organization is completing relatively more projects (with the same resources) than was initially assumed it was capable.

Necessary Assumptions

In the planning [process] the time estimates of the tasks on the Virtual-Drum were set to reflect the capability of the Company to finish projects (this rate is not dictated by capacity but by the level of synchronization). The planning mechanism controls the release of new projects according to the Virtual-Drum. The rate at which the Company finishes projects cannot be (constantly) higher than rate of release. The improved execution (steps 5:114.2-4 and 4:11.5) increases significantly the level of synchronization and therefore the capability of the Company to finish projects at a higher rate. Therefore, if in the planning, the Virtual-Drum "loading" will not be (frequently) adjusted to reflect the improved synchronization, the Company will finish projects not [at] the rate it is

capable of but [at] the historic rate it had when the planning was initiated.

The first and most important thing to remember when considering adjustments to the Virtual Drum is that except in very rare cases adjustments should be positive, i.e. increasing the rate at which the organization releases and completes projects. This is why we set the initial stagger by emulating the existing Virtual Drum – matching as closely as possible the actual rate of project completion at the time. It is logical to assume that the organization will get better, so there is very little risk associated with beginning to stagger by emulating the existing Virtual Drum.

On the other hand, it is very dangerous to be too aggressive in the initial pace of the Virtual Drum. If the organization is unable to keep up with the pace, projects will go red and then black at an increasing rate, and the organization will have to reduce the pace of the Virtual Drum, meaning projects will be delayed and customers will have to be delivered the disappointing news. Note that this step itself assumes both explicitly and implicitly that the Virtual Drum adjustments will be positive.

Even as you improve in your execution of individual projects, improvement in the overall performance of the project portfolio will be harder to achieve if these positive adjustments are not made to the Virtual Drum at reasonable intervals. In other words, as the Necessary Assumption states, "if in the planning, the Virtual-Drum "loading" will not be (frequently) adjusted to reflect the improved synchronization, the Company will finish projects not [at] the rate it is capable of, but [at] the historic rate it had when the planning was initiated."

Strategy

The Company's rate of completing projects is in accordance with the Company's changing capabilities.

In order to make sure the organization is always operating at its full potential, the rate of project completion will match the company's capabilities.

Parallel Assumptions

When synchronization improves and the rate at which projects are released [has] not been increased, the number of projects waiting to enter the chosen integration will drop. After some time the number of projects in the chosen integration will drop (from time to time) below the restricted number allowed. The rate of drop in the number of projects in and before the chosen integration, is an excellent indicator to the required adjustment in the rate of the Virtual-Drum (adjusting the time estimates of the tasks on the Virtual Drum to reflect a rate of execution to be higher by the corresponding rate of drop in number of projects). Check: In such a situation, there is no significant accumulation of projects in other stages - which is the case when there is a sharp deterioration in the Company's capabilities (due to losing too many people or due to a significant shift to projects that require a still not fully-mastered technology).

Fortunately, there is an easy way to know when the right time might be to make adjustments to the Virtual Drum. Our first indication is when the number of projects being worked on in integration falls below the preset number allowed. Once again, a single data point, or this situation occurring only once, is not sufficient for making a change. However, if you see the number of projects in integration periodically dropping below the allowed amount, it's time to take a closer look. Therefore we need to track the actual number of projects in and before the integration area for each portfolio and watch for trends.

The Parallel Assumptions contain a warning, however. Before making an adjustment, we need to ensure that the drop in projects within the integration area is not due to a reason other than improved performance. So you will want to check to see that new bottlenecks have not arisen in your system, or in other words you are not seeing the accumulation of projects elsewhere. For example, if you have lost skilled workforce upstream of the integration area, you will see an accumulation of projects upstream, while at the same time seeing the number of projects in integration drop. In a case like this adjusting to a faster pace would be a big mistake, because projects would only accumulate at a greater rate, throughput at the Virtual Drum would not increase, and in fact would probably decrease due to the increased workload and multitasking at the bottleneck.

We should also make sure that we have not experienced a drop in confirmed sales – in fact our new reliability should be reflected in an increase in sales.

Tactics

The Company constantly monitors the number of projects in and before the chosen integration and periodically adjusts the rate of the Virtual-Drum in accordance.

Once you have eliminated other reasons for the drop in the number of projects in and before the chosen integration area, you can increase the rate of the virtual drum. As a sanity-check before making the change, you should seek a consensus of workers and management that you actually have improved enough to warrant setting new expectations and that no other problems exist. Another good indicator is that the number of green projects is increasing over time. At this point, there are a number of ways to increase the pace of the virtual drum, but the safest seems to be to do it by reducing the size of the capacity buffer. This will pull projects closer together, and pull the delivery date of all projects upstream of virtual drum start a bit to the "left," or earlier in time. Alternatively, you could reduce the task duration estimates for tasks in the Virtual Drum phase. Both methods will accomplish the same thing. But before you quote new, earlier delivery dates to the customer, if you wish to do so at all, you should allow some time to be sure the system can smoothly operate at the increased pace.[64] If it can, it's good news for everybody.

NOTE: if you have significantly underrated your capability, and you are getting frequent and increasing "drops" in work load on the Virtual Drum, you may also consider increasing the number of projects allowed in integration. However, you should be conservative and again allow some time to make sure you are not overloading your system in any way.

---

[64] There may be other good reasons for not immediately (if ever) informing your customers of your shortened lead time especially when you reliability alone is attracting more business than you can handle. For example, see 3.2.1 "Shift to Bonus Deals" and Level 4 below it in the Projects S&T tree (not covered in this book).

Caution: this is a good time to remind you to make sure your salesforce is prepared[65] to bring in more work if necessary, or the rate of your increase in sales might not keep up with the increase in the pace of the system. Increasing sales is highly preferable to allowing gaps to appear in the pipeline, and dealing with the pressure to reduce manpower, which can also reduce your precious protective capacity, and send you back into an overloaded condition.

Was the step correctly implemented?

The test for whether this step was correctly implemented takes longer to realize, since adjustments to the Virtual Drum usually begin to take place sometime (weeks or months) after the launch of the CCPM system. First, make sure your initial stagger is not too aggressive – it's much easier to add projects into work than having to slow things down and suffer the consequences. Next, you want to make sure you are monitoring the number of projects in and before the chosen integration area, watching for a negative trend in drops. Once the decision is made that your productivity is outpacing your schedule (the stagger), you will want to make sure the adjustment is properly made and followed up on. Then the cycle repeats anew.

Were the expected effects realized?

The expected effects include an almost immediate increase in throughput through the Virtual Drum (and subsequently in project completions), and the increased ability to introduce new projects into the system. If these effects are not seen in the near-term, check your assumptions!

---

[65] See 3.1.2 "Reliability Selling" and Levels 4 and 5 below (also not covered in this book).

What mechanism has been put in place to ensure compliance with the step over time?

A Virtual Drum Adjustment policy for your organization is a must. It should be accompanied by a log that records every adjustment and the relevant parameters to track effectiveness – including screen shots of fever charts.

You will want a mechanism that monitors and reports the number of projects in, before, and after the chosen integration area. Projects in the integration area should also measure deviation from the target. When your improvement indicates a trend of drops in the number of active projects available, you will want a mechanism (again, possibly policy-related), that adequately follows up on the change and verifies an increase in flow.

You may also wish to measure days between each integration area completion and days between each project completion and watch for trends.

Section Six: The Final Touches

Flow is the number one consideration

Chapter 25 – 4.11.5 Mitigating Client's Disruptions

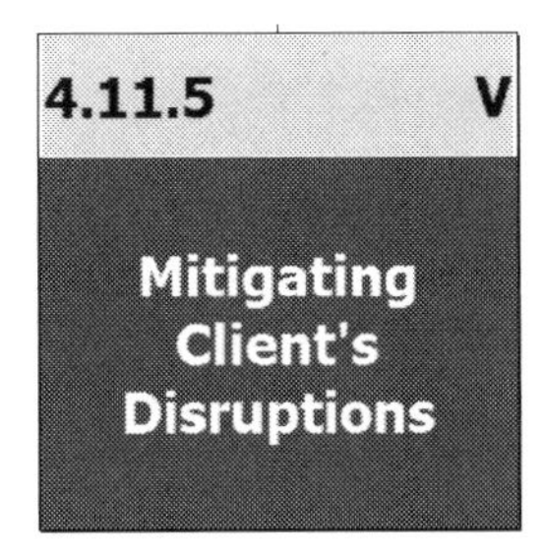

In Chapter 14 – 5.112.3 Worried Clients, it was stated that many times we are our own worst enemy – up to 80% of all supplier delays can be traced back, at least in part, to our own actions/lack of action. The same can be held true to a lesser extent for delays perceived to be originating with our clients. However, it is very common for delays to genuinely originate with the customer. The reason for this is our customer companies are often quite similar to our own. They have the same problems and are guilty of many of the same actions/lack of actions as were outlined at the end of Chapter 14.

Necessary Assumptions

Many times the client is the cause for the project being late by delivering late on inputs (information, components, authorizations etc.) and/or by demanding specification changes.

The truisms contained in the Necessary Assumption are all too common for many of us. Yet now we at least have the ability to understand why client delays occur. Maybe they need to implement TOC! Unfortunately, delays caused by late customer inputs are usually absorbed by the project without changing the delivery date. This is usually because of an assumption held by salespeople and client representatives that certain efficiencies can be found that will eliminate the need for changing the due date. This mindset tends to prevail even when the delays involve specification changes. And the customer will rarely blame himself for a late project caused by his own input delays and earlier specification changes. Therefore in order to mitigate the damage caused by client delays, a strategy is needed.

Strategy

The Company has very high due-date performance even in cases where client inputs are required and/or specification changes occur.

The verbalization of this strategy acknowledges that client delays are a real problem, which in itself is an important step. This allows for a proactive approach for dealing with client-originated delays rather than only a reactive approach. A look at the Parallel Assumptions will help us formulate tactics for mitigating client delays.

Parallel Assumptions

CCPM provides the ability to identify which delays in input and which specification changes are likely to delay the completion of the project (most input and specification changes do not delay the completion of the project, however, they are still extensively used as excuses). When the client professionals realize the quantifiable impact of their actions (delayed input or specification changes) they are very likely to change their behavior in accordance.

One of the great strengths of Critical Chain as compared with other methodologies, is its ability to identify and quantify the impact of delays even very early in a project. In many cases delays such as these actually do not threaten on-time project delivery. Since truth should always be a high value in a TOC organization, this is good news and minimizes the temptation to use such delays as excuses for our own mismanagement. However, some client delays do legitimately impact project delivery. And when this is the case, it is important that the client be made aware.

We have already trumpeted the benefits of increased communication, as well as the sifting out of truly important things from those not so important. Just as within the organization, good information can be extremely valuable when shared outside the organization.

Whenever a client's delay will endanger a project's delivery date, the information should be shared with the client as soon as possible. If the client believes the information is reliable, it is likely that the client himself will take action to mitigate the delay. The client may even offer (can you imagine it?) to move the delivery date later in time.

Tactics

The client professionals are exposed to the CCPM project network and the logic of its buffers. The Company people who interact with the client are professional at communicating the impact the client actions have on the completion of their project and the resulting damage. The mechanism is in place to adjust due-date commitments when applicable.

Some of the paradigm shifts involved with CCPM pertain to the salespeople. These shifts require that our salespeople are educated in Critical Chain theory and in the projects S&T tree. We have already discussed the new paradigm of quoting delivery dates based upon when a project can be inserted into the pipeline in accordance with its stagger. In this step, salespeople should be able to communicate our entire new methodology to our customers. Buffers are not intuitive and must be explained to clients before implemented, lest they ask (or even demand) for them to be taken away. At the same time, buffers can be a very powerful sales advantage as CCPM enables full transparency with clients – and they love it when we give them access to their project data and buffer consumption rate.

This communication allows for 1) the customer understanding our behavior, 2) the customer gaining confidence in us as a supplier, 3) the customer contributing to his own success, and 4) the customer even possibly wanting to explore the benefits of TOC for his own organization.

When a customer delay will endanger a project's delivery date, the customer should be shown the buffers and the extent of damage the delay is expected to incur in the project. This is both more accurate, more helpful, and more equitable for the client then simply insisting on something like, "a day-for-day slide" to a project's delivery date for each and every client delay. Interactions by the salespeople (or others) with the client should always be professional, respectful, and non-accusatory. There is no place for blame in TOC. We simply need to work together to achieve better results.

The last part of the tactic says we need a mechanism for adjusting delivery dates. Therefore we have to use our common sense and keep in mind accepting significant changes to one project can have an affect other projects in the stagger, and we need a systemic approach that maintains flow as the number one consideration.

Chapter 26 – 4.11.6 Managing Sub-Contractors or Contracted Sub-Projects

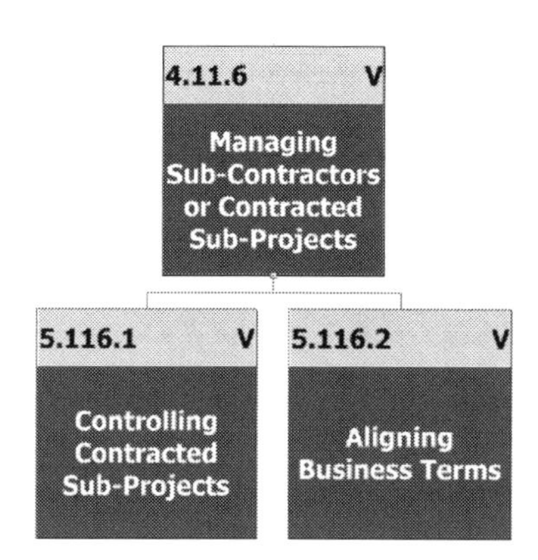

We now come to the last Level 4 entity in *The Critical Chain Implementation Handbook*, which continues along the theme of communication. Because there will always be suppliers, customers, and others outside the organization which will be involved in our projects, any guide on implementation must address relationships among the parties. It is simply a fact of life that entities outside our direct control can play a big part in our delivery performance, and therefore our reliability and reputation.

Necessary Assumptions

Almost all multi-project environments suffer from notoriously bad due-date performance. In cases where a sub-project is contracted, the more the Company improves its performance the higher the likelihood that the prime reason for a delay in a project completion will be delays caused by the sub-contractor.

It is very common for projects organizations to use subcontractors, sometimes for major portions of work within their projects, and sometimes using completed subcontractor projects as inputs to their own.

Just as many of our customers are also projects organizations, many of our suppliers and subcontractors can also be projects organizations. And as you are fully aware by now, projects organizations in general "suffer from notoriously bad due-date performance."

This is bad enough when we too are struggling, but our own shortcomings tend to mask the shortcomings of others. When we significantly improve our operations by the application of Critical Chain and the Theory of Constraints, the poor performance of others becomes glaring. As we are improving, the major cause for delays shifts more and more to those outside our organization. Soon we begin to realize that dealing with these external forces is crucial to our own success (and our reliability). Therefore it behooves us to have a set of strategy and tactics to deal with subcontractors and subcontracted projects.

Strategy

The Company has very high due-date performance even in cases where sub-projects are contracted.

As with many of the strategy statements, the initial response of some may be, "You're dreaming!" But this feeling can be reduced when you consider two things: first, that strategy is only "what" we want to achieve and not "how" we plan to achieve it; and second, that any lesser strategy would be a concession that we really don't believe we can succeed. In spite of the fact that we have little direct control over the operations of our subcontractors, we can still take actions (tactics) to help subcontractors contribute to our success. What should these actions be? To find them, we need to look at the conditions that exist in reality – or what we call Parallel Assumptions.

Parallel Assumptions

CCPM provides the ability to zoom in on the task that jeopardizes the completion of a project. The relationship between a prime contractor and a sub-contractor enables the Company to provide the needed focusing for its subcontractors. When the business relationships with the sub-contractor are hourly based, a prime concern of a sub-contractor is to "click" enough hours. Even in that case, it is possible to use the eagerness of the contractor to get higher fees per hour to remove the conflict with the Company's on-time needs.

Since this is an important step that has a lot of impact on our reliability, let's take a close look at the Parallel Assumptions one at a time:

1) CCPM provides the ability to zoom in on the task that jeopardizes the completion of a project.

Most projects have multiple paths that at some point converge before the project can be called complete. Ideally we want the Critical Chain itself to be driving buffer penetration. However, it is possible that delays in feeding chains can delay the Critical Chain. This means any chain can potentially become the "most penetrating chain," and that chain's current active task is the "most penetrating task." CCPM provides the ability to identify this task, providing focus for the organization. Sometimes the task that jeopardizes the on-time completion of the project is a task (or sub-project) which is carried out by a sub-contractor. Sometimes the sub-contractor task or sub-project is in the future, but has to be considered when formulating buffer recovery plans. And sometimes you just have a general concern that a sub-contractor will be available and ready to start executing as soon as the schedule calls for it.

2) The relationship between a prime contractor and a sub-contractor enables the Company to provide the needed focusing for its sub-contractors.

Usually the relationship between a prime contractor and sub-contractor is a *contractual* relationship. This means terms and conditions can be written into a contract that the sub-contractor is bound to follow. The contractual vehicle can help create focusing for sub-contractors. But *relationship* means more than just words written in a contract. If a relationship is adversarial, or is perceived by the sub-contractor to be something akin to a master-slave relationship, cooperation can suffer greatly, and one side or the other may push the relationship into one of win-lose. TOC is about creating win-win relationships.[66] Win-win relationships foster cooperation and contribute to an organization first becoming reliable, and then ever-flourishing.

3) When the business relationships with the sub-contractor are hourly based, a prime concern of a sub-contractor is to "click" enough hours. Even in that case, it is possible to use the eagerness of the contractor to get higher fees per hour to remove the conflict with the Company's on-time needs.

One thing that must be remembered is that the prime contractor and the sub-contractor share one very important objective. They both want to *make money*. In fact, they both likely want to make as much money as possible. This means that sub-contractors will, as a rule, make decisions that are in their own best interest. As a general statement, they will *go where the money is.*[67]

In addition to TOC being about win-win relationships, it is about *holistic solutions*. This means the entire company (hopefully) is pulling together to follow and meet the goals of the S&T tree. "Entire company" includes every department, and in this case, it includes the purchasing (and legal) departments, or the departments responsible for securing, reviewing and paying sub-contractors. Unfortunately, many departments still operate, at least in part, as "silos." To make matters worse, silos often try to optimize themselves without regard to the holistic organization. Worst of all, this sub-optimization usually puts a significant amount of energy into cost-savings.[68]

---

[66] Win-win relationships are not based on compromise. Compromise is actually lose-lose, since both sides must give up something they want, and neither are necessarily happy about it.

[67] This may mean money now, or money in the future. Often promises of future business can be as good as of incentive as money paid today.

[68] Many lucrative opportunities exist in the space between silos. Unfortunately, these can only be accessed through collaboration between the silos.

Therefore it is very common that the number one consideration of purchasing departments is finding the lowest possible price. At times this objective can even be brutal. You have probably heard the horror stories about big companies that routinely "make" and "break" smaller companies. Mom and Pop operations hit the jackpot when contracted by a big company. Their fortunes, and their size, explode overnight. But very often it isn't long before the big company starts demanding price cuts. Sometimes they demand a certain percentage price cut on a periodic basis, such as yearly. Now the previously small company has a real dilemma. Losing the contract to a competitor means disaster, especially when the small company relies on this single customer, or a very small number of customers. On the other hand, accepting the continual price cuts also threatens the survival of the company, as productivity increases are sometimes surpassed by increases in operating costs. The celebrating that began the relationship turns into despair.

This may work for the big company for a while, but sooner or later it will come back to bite them. This is because there are considerations other than price. The big company also needs reliability, but when the company is seen as a slave master pushing low price over anything else, reliability takes a back seat every time. If cost is the number one consideration, it means flow is *not* the number one consideration.

Of course cost is a legitimate consideration, and there is nothing wrong with trying to find a good price. But if flow is not the **number one** consideration, the company can find out it has injured itself sacrificing delivery performance (which can be big money) for a small cost savings. It is always a long-term positive to have a good relationship with your suppliers – especially your key suppliers. And in certain situations, good relationships can be utilized to structure ad hoc win-win solutions aiding buffer recovery.

Even the department responsible for sub-contractor relationships should keep flow as the number one consideration, be aware of buffer status, and contribute to buffer recovery.[69]

---

[69] The Full Kit Manager can also play an important role in ensuring this.

Tactics

The Company provides on-going focus to the sub-contractors. When appropriate, the Company is careful to provide the right incentives for satisfactory on-time performance to its sub-contractors.

We will use our contractual relationship with sub-contractors to help align their operations with the needs of our organization. In addition we will foster cooperation and win-win solutions both in our general relationship and selectively with additional incentives in situations that require special action and attention.

Sufficiency Assumptions

Many times, things not under the Company's direct control are still under the Company's (strong) influence.

Even when we understand this logic, we realize that we don't yet have all the information we need to manage sub-contractors and sub-contracted projects. The Sufficiency assumption alerts us that more detail exists at Level 5. But in this case the Sufficiency Assumption also provides us with hope. Even though our sub-contractors are not our employees, there are still positive things we can do to solicit (and manage) their cooperation and contribution to our success.

Chapter 27 – 5.116.1 Controlling Contracted Sub-Projects

The first scenario we will consider is one where a subcontractor produces a project that is used as an input to our project. For example, this could be something like a fuselage assembly for an aircraft, or a large piece of industrial equipment that is used as a component in the construction of an oil refinery. It is an entire, separate project. Sometimes all work can be done on the subcontractor's premises. Sometimes the work is done on the prime contractor's premises, but it is still a separate project. Sometimes the work can be done in both locations by subcontractor employees, but again, it is still a separate project.

Necessary Assumptions

Sub-contractors for sub-projects are usually companies whose environment is a multi-project environment. In most multi-project environments due-date performance is notoriously bad in spite of management's determination to improve.

Usually subcontractors contract to more than one client, or are at least working on more than one project at any given time. This means they, like you, are multi-project organizations, with all the same problems and pressures working against their on-time delivery. You may press them to improve, and they may make their best effort, but unless they go through the same paradigm shifts as you and change behaviors, their delivery performance will remain spotty at best. Understanding that pressure alone will not eliminate damage done to your organization, it is best to have a set of strategy and tactics to deal with subcontractors.

Strategy

The Company helps its sub-contractors to better deliver on (or before) time.

In too many cases, relationships between contractors and subcontractors are considered as little more than blind business arrangements. Contracts call out specifications and delivery dates, and subcontractors are expected to handle all of the necessary and sufficient details to produce the deliverables. Inputs from the contractor company are perceived by the subcontractor to be generally negative. Either the contractor is changing his mind about something and injecting chaos into the subcontractors operations, or he is making unreasonable demands.

With TOC, a paradigm shift is required for these relationships if the contractor is to expect greatly improved delivery performance from his subcontractors. This is a change from a blind business relationship to one of help and cooperation. But how exactly can a contractor help his subcontractors help him? And how can he do it in a way that is not seen as an additional burden for the subcontractors?

Parallel Assumptions

It is common for a prime contractor in project environments to dictate to its sub-contractors some procedures of reporting. The Company can demand that the sub-contractor will supply a PERT network for the contracted sub-project. The Company can, for their own purposes only, change the supplied PERT to a buffered PERT by removing 50% of the task estimates and placing them as buffers at the end of the applicable paths (there is no point in attempting to remove resource contingency). The company can demand that the sub-contractor will report daily his progress (remaining duration) on the network tasks and use this report to identify the task that currently endangers the on-time performance of the contracted sub-project. Regarding the "endangering task" the Company can demand a report on the recovery actions. Experience shows that the above procedure, accompanied by proper explanation of the underlying mechanism, is appreciated by sub-contractors.

In an ideal world, your subcontractors would, like you, be users of Critical Chain. They would be dependable, and reporting would be easy. For all intents and purposes they could be treated as task managers – daily reporting the remaining duration of their entire project. You could depend on this information – it would be meaningful because it would be informed by their own buffer penetration, and it would reflect accurately in your project buffer penetration percentage. But in the real world this is rarely the case. Prime contractors and subcontractors are separate companies with separate management systems. Their internal workings are often more mysterious to one another than not. Too often, all you have been able to do until now is either take their word on when their projects will be complete, or else try to micromanage them in some way.

But there is an alternative. The contractual vehicle between your companies can be used to secure some of the benefits you would get if your subcontractors were on CCPM. But since you probably don't want to be a micromanager (don't you have enough to do already?), you should consider the following carefully and think it through before jumping to put further requirements on your subcontractors.

Increased communication is always a wise approach, and it makes good sense to educate your subcontractors to some extent about CCPM. Maybe it will pique their interest in adopting CCPM themselves.[70] At least they will understand why they are being asked to take these steps, and if you do it correctly, you may find they appreciate the process.

One of the things you can do is to contractually require your subcontractors to submit PERTs of their projects. But the purpose is not micromanagement – it is information. Their PERTs may not be perfect, but the information they provide is superior to information received if their projects are only "black boxes" to you. If you do not require a PERT, but instead only ask a subcontractor to report remaining duration on the entire subproject, the information you receive will be of little use as a forward-looking mechanism. For example, if their project is scheduled for 100 days and each day you ask for remaining duration, you may very well only get a "countdown" – 99 days, 98 days, 97 days, etc. This provides no

---

[70] Some might endorse requiring subcontractors to use CCPM, but this is not always practical.

information as to whether the project is healthy or in danger, or to whether it will actually be delivered on time. Having the subcontractor break the project down into tasks and submit it as a PERT however, and to provide remaining duration by task, can give you much better insight into their projects' health. But they don't have to go to minute detail - some granularity is better than none. You also don't have to ask for resource information. That's their business, and it's not practical for you to try to resolve your supplier's resource contention.[71]

Internally, you can take a subcontractor's PERT and transform it into a Critical Chain project plan. You can remove safety from their tasks by cutting task durations in half, and then you can place the aggregated safety at the end of each chain as buffers. Please note the difference here, however. Don't cut the buffers in half as you would if it were your project. Keep all the safety and put it at the end of the chains. If the result (project plus buffer) does not exactly match the total duration they have committed, you can adjust the buffers until it does.

Now make sure you are asking for remaining duration at the task level they have defined – for the current task they are working on. This provides several benefits:

1) You get early warning for project problems' impact on the buffers.
2) You can offer help when appropriate, especially if it is *you* who is the source of the delay. Maybe they need something from you, such as a drawing, or an approval (See Chapter 14). Maybe you can help them get an important material, etc.
3) You will only ask for the subcontractor to take action when the need is legitimate (i.e. their buffer is red).
4) You will be seen by the subcontractor as more of a partner than a master.
5) You can better ask for and monitor the subcontractor's recovery plans.[72]

---

[71] Experience shows that it is likely you will have to do some remedial work on suppliers' PERTs, so why not, as part of the contract, insist on holding a one-day planning session with you where you cover the basics of network construction, backward planning, proper task definition, and optimal assignment of resources.

[72] All of these benefits are amplified by holding a planning event (see footnote 71)

Tactics

The Company demands that the sub-contractors submit the relevant PERTs and daily reports on tasks' progress. This data is used to provide ongoing focus to the sub-contractors.

Using the information provided in the Parallel Assumptions section, write the appropriate language into future subcontractor contracts. Educate your subcontractors on the purpose of your new requirements. Provide focus, help, and guidance as necessary. You may also consider hosting their CCPM schedule in your system and providing access to them free of charge.

Was the step correctly implemented?

The actual execution of the step varies according to the process you adopt. Carefully consider the process, educate and discuss it with your subcontractors. Develop your own criteria for correct implementation. An Auditor can review your criteria and make suggestions if necessary. The Auditor can also determine compliance with the criteria. However, one question common to all processes should be: Are key subcontractors of major subprojects all required to submit PERTs and provide daily updates?

Were the expected effects realized?

The expected effects were listed as benefits in the Parallel Assumptions section above, which also contain benefits for your suppliers: Early warning for impact on your buffers; the ability to offer help when appropriate; prompt supply of something they need from you, such as a drawing, or an approval; requesting action only when the need is legitimate; being seen by the subcontractor as a partner; and better access to the subcontractor's recovery plans.

What mechanism has been put in place to ensure compliance with the step over time?

A new or modified internal policy document, (with appropriate follow-ups) for managing subcontractor relationships is necessary for providing guidance for future negotiations.

Purchasing personnel should be directed and measured on helping project teams steer work toward those subcontractors who are most willing to provide PERTs and otherwise cooperate with the tactics laid out in this chapter.

Chapter 28 – 5.116.2 Aligning Business Terms

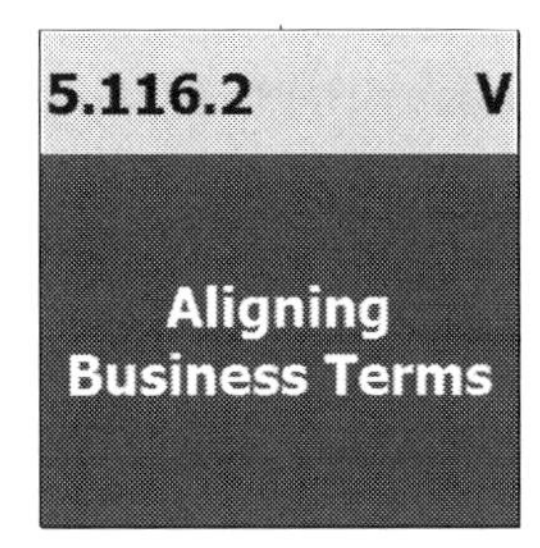

Although subcontractor projects (sub-projects) can have an hourly component as part of the arrangement, it is even more common for hourly-based compensation to be paid to subcontractor companies when the subcontractors are providing labor for *your* projects. In either case, whenever hourly compensation is involved it has its own set of ramifications, and to be successful it behooves the company and its subcontractors to be on the same page.

Necessary Assumptions

In cases where the business relationships with the sub-contractor are hourly based there is an inherent misalignment between the Company and the sub-contractor (the Company wants to get the deliverable on time and with as few hours as possible while the sub-contractor wants to work as many hours as feasible). In such cases incentives for early completion have a limited (if any) effect.

Whenever the goals and objectives of two separate entities are not 100% in harmony, "misalignment" occurs. Therefore, misalignment can occur when ever two separate business entities seek to maximize their own profitability while in relationship with one another.

While there is nothing wrong with each company looking after its own best interests and seeking maximum profit, it is extremely easy for such relationships to be viewed through a win-lose lens. This means one or both companies will most likely not get everything they want from the relationship.

Due to laws concerning job security and benefits for direct employees in India, project labor tends to come heavily from subcontractor companies. Sometimes the prime contractor only supplies management, and subcontractors perform all project labor. This situation is effective for highlighting the misalignment that can occur between prime contractors and subcontractors.

Obviously, both prime contractor and subcontractor companies want to make as much money as they possibly can. Prime contractors want work performed in as few hours as possible, while subcontractors want to log as many hours as possible – hence the misalignment is clear. Prime contractors want to minimize hours paid, while subcontractors want to maximize hours worked. This alone produces conflicts. But it can go much further.

When a prime contractor project is delayed for reasons unrelated to the subcontractor, the prime contractor must ask the subcontractor to wait, potentially with an idle workforce, while the problem is resolved. Then the subcontractor is expected to immediately resume work on the project. But from the subcontractor's perspective, this is a terrible arrangement.

In order to secure cash flow, the subcontractor may take work on another project for another contractor. When the original contractor's problem is resolved, the labor force is not available. Therefore it is the prime contractor who has to wait, or find alternative labor, which could have its own set of problems.

In addition, the subcontractor will want to be sure that he does not finish early, because in doing so he may not be able to log the maximum number of negotiated hours. Of course, this is all made worse when the number one consideration of the purchasing department of the prime contractor is cost savings. This is the epitome of misalignment. We need a good set of strategy and tactics to overcome this misalignment.

Strategy

The Company's sub-contractors are effectively incentivized to deliver on, or before, time.

From now on, it will be the policy of the company, and of its purchasing department, to offer effective ad hoc incentives to subcontractors when the situation warrants. In addition, as long as quality and safety are maintained, a subcontractor should be incentivized to finish a contract on or ahead of schedule.

Parallel Assumptions

For sub-contractors who are used to hourly-based contracts, the prime concern is fee-per-hour. Usually, the bigger the delay in a project the higher the number of hours invested. An incentive that offers a sub-contractor a higher fee-per-hour for early delivery is effective in inducing the contractor to deliver earlier. Experience shows that, in many cases, the higher fee-per-hour was more than compensated by the reduction in number of hours invested.

We want to avoid "busy-work" by the subcontractor, as well as to remove any incentive to "find" reasons for delay or to "stretch out" work by the subcontractor. At the same time, if the incentive offered is too small, it will be ineffective. Experience reveals that raising the fee-per-hour by sufficient amounts for each increment of early delivery will effectively drive the behavior desired. This provides a win-win for both companies, since the subcontractor is able to maximize his overall profitability, while at the same time, the higher cost of labor for a small portion of a project is usually easily offset by the increased throughput[73] of the prime contractor through earlier deliveries and shorter lead times.

In certain cases where the availability of the subcontractor is in question, a guarantee of payment or premium for availability may be appropriate incentives.

Tactics

[73] There is more to be considered than simply the cost savings available to the company for one specific project. Shorter projects overall means more projects can be done by the same resources in the same timeframe. Therefore the cost of the premium should be considered in light of the overall throughput of the company.

The Company offers significant bonus payments (per hour) to its subcontractors for on-time (or earlier) delivery.

The tactic itself is straightforward, although the necessary resources should be utilized to determine what the appropriate amount of an effective bonus would be. To get the best effect and drive the right behaviors, the bonus should be paid on an hourly basis, rather than as a lump sum. Subcontractor planning workshops with experienced internal staff are key before offering incentives.

Was the step correctly implemented?

Were incentives written into the original contract or added by amendment? Were the terms of the incentives satisfactory to the subcontractor for the given situation? Was payment for the subcontractor's performance prompt and made with the proper positive (and grateful) attitude?

Were the expected effects realized?

The expected results of this step are the evidence of desired behaviors (availability, no superfluous logging of hours), and especially early completion of subcontractor work. We also expect to see a reduction in the number of change orders.

What mechanism has been put in place to ensure compliance with the step over time?

You may wish to document a policy of reviewing project networks prior to project release to identify the areas where subcontractors will be needed, and what the appropriate incentives for these areas should be. You also may wish to monitor whether the incentives planned were actually offered and paid, and to document any unexpected issues or surprises in the process.

Chapter 29 – A New Candidate for the S&T Tree – Freezing Tasks at the Department Level

In this chapter I'd like to introduce an action (tactic) that might be considered as a future addition to the Projects S&T tree. It is a step that should be used only as necessary, and therefore in a minority of cases, but nevertheless it might be a good idea to document it in the tree so people are aware of it and understand the logic. Dr. Goldratt, always the true scientist, had us test the step at a client in 2010, ensuring the hypothesis was sound before making it public. In that particular case, this step was included under "Managing Execution," inserted just before Virtual Drum Adjustments. The company reported a significant positive effect from applying the step – task switching decreased and projects were accelerated. Based on a conversation I had with Dr. Goldratt concerning the success, I understood that he was considering adding the step to the generic Projects S&T template.

Realizing the damage that task switching causes in an organization – that even one department which suffers from bad multi-tasking can radiate significant delays into many projects, a necessary assumption for further eliminating bad multi-tasking might be:

Even after freezing of projects and preventing early release of project legs, some resources may find themselves under pressure to frequently switch between tasks; there might still be departments where prolific bad multi-tasking still exists.

In certain situations, even after reducing load on the system by a significant amount (greater than 25%), a small number of departments might still experience too much task switching. In such a case it might not make sense (or the damage to reputation might be too great) to freeze projects a second time. Therefore we need a strategy aimed at eliminating the problem:

The company eliminates bad multi-tasking to the greatest possible extent.

In 5.111.1, Prioritizing and Freezing (Chapter 7), we attacked task switching by reducing the number of projects in work and controlling the load at the new lower level. In most cases this approach is sufficient to minimize task switching to the point that it produces minimal damage.[74] This step is more specific, and further targets task switching, explicitly stating that we will do everything we can to not minimize, but eliminate task switching. We may never achieve perfection, but it will not be from lack of effort.

Eliminating task switching is a noble strategy. But how do we achieve it? It seems clear that we need some tactics, and to get there we need to develop some Parallel Assumptions. Sometimes Parallel Assumptions include actual methods we can use to help us arrive at tactics. Let me suggest some thinking that might be included in leading us from the strategy of eliminating task switching to one or more appropriate tactics. I have not proposed official Parallel Assumptions here, but rather I am telling part of the story concerning how we went about testing the hypothesis of freezing tasks by department

To determine the severity of bad multi-tasking within departments, Dr. Goldratt proposed a simple test. Conduct a survey of the people actually doing the work in your projects. Goldratt insisted it must be the workers, not their supervisors, because otherwise we won't get honest answers. Ask the workers this question: "In the past month, how many times have you been asked to stop work on one task in order to begin working on another task?" Collect and average their answers. If the number is greater than the maximum number that you determine is allowable for your environment (at our client, Dr. Goldratt suggested we use "3"), there is too much bad multi-tasking.

At this point some of you may be laughing. "Three? In a month? I'm asked to switch tasks three times in a day!" This is not uncommon. But switching tasks at such a frequency indicates severe, chronic bad multi-tasking. At our maximum allowable number of 3, if the answer from our client was 5 or 6 times a month, a 25% reduction might have been

---

[74] Not that the goal is minimal damage. You should always strive to eliminate task switching, and watch for negative trends. But it's a good start – most often good enough for a successful initial implementation. This step just makes elimination explicit and proposes a solid action to help you get there.

sufficient. But the answer (varying by department) was 20, 30, or more, so it is clear that more than 25% should be frozen. We let the client decide the reduction for himself by department. The actual reduction they chose was between 60% and 80%!

A knowledgeable team should try to come up with a percentage that will reduce the level of task switching to the maximum allowable number (or fewer) per month.[75]

Now let's move on to the "how" to freeze tasks by department. The following are suggested tactics for doing so:

1) ***Departments which suffer from bad multi-tasking are instructed to freeze tasks. Given the information gathered, and based on the total average number of active tasks in the department, decide what percentage to freeze.***

Another important piece of information you should collect is average task duration time. Compare the ratio of the number of disruptions in the month to the length of average task duration. For example, 10 disruptions per month where average task duration two weeks is far more damaging than 10 disruptions where the average task duration is one day.

Remember that in this case we are freezing *tasks* by department, not projects across the entire organization. Based upon the degree of bad multi-tasking discovered, decide what overall percentage of tasks should be frozen. The department head should not be told *what* (which tasks) to freeze (he knows this better than we can), only how much to freeze. (This number is the minimum percentage that should be frozen. The department itself may realize that even a higher percentage of tasks can be frozen.)

2) ***Implement the freeze.***
3) ***Verify that freezing has been properly carried out.*** This means not only confirming that the right percentage of tasks were frozen, but also that no "unofficial" or "informal" work is being done on the

---

[75] The proper number could be 35%, 50%, even 70% or 80% or more! The important thing to remember is the percentage chosen must be sufficient to maximize flow and throughput (more tasks completed) through the department.

frozen tasks. In other words – no cheating. Check for expected effects.

4) ***Follow with (task) acceleration, defrost, and introducing new (tasks) procedures, in order. A general good rule for defrosting is release of a new task is allowed only when a task has been delivered to the next department.***

This process will rapidly free up capacity and increase the velocity of task completion in each selected department.

Chapter 30 – Closing Comments

Did I mention flow was the number one consideration?

As we reach the closing chapter of this book and the implementation journey we have taken together, let's pause and review the paradigm shifts an organization must complete for a successful implementation of CCPM in a multi-project environment. We will also cover how these paradigm shifts, although not easily implemented and "lived", are the enablers of the Decisive Competitive Edge a company can attain when CCPM is correctly implemented.

We will look at the paradigm shifts in two steps. First I want to highlight the foundational, holistic paradigm shifts that TOC in general, and a successful CCPM implementation in particular require. Then we shall review specifically each entity of the S&T covered in this book and emphasize the relevant paradigm shifts.

The number one foundational paradigm shift is of course: **Flow is the number one consideration.** If you really look deeply into your organization you will find that this imperative contradicts most, if not all of our measurements today, and many of what are considered the "right" actions by functions and functional heads. From local efficiencies to top management policies, measured against this paradigm you will be amazed by the contrast between what is (our current driving considerations, such as cost reduction) and what should be.

Using flow as the compass, you can take each and every local decision and ask a simple question: Is this action/decision I am considering going to contribute to increased flow, or is it in reality going to slow the flow/project down?

The second paradigm shift and core difference between CCPM and traditional project management is moving from the idea that to "finish a project on time we must complete each task on time", to the complete opposite approach: The only consideration is to finish projects on time – **while individual tasks can be late we can still finish our projects on time if there is sufficient buffer to protect them from variability.**

The practical meaning of this shift is we remove safety from tasks and protect the project and feeding chains with sufficient buffers. Then, for multi-project environments, we sync all projects on the Virtual Drum, choke the release of new projects, start legs when they are due to start, etc. It also means we stop asking task managers for completion percentages, and stop holding them accountable for making daily progress on each and every task. Instead we ask them for expected remaining duration of tasks, focus on tasks in priority order, and according to the impact on buffers make decisions on how to recover buffer and complete our projects on time.

But to have the above abilities in an organization yet another paradigm shift is needed: **Throw blame out the window**. Instead of asking task managers "why are you late?" and placing blame on them for not doing well, we take a completely different approach and ask a different question: "How can I help you go faster?" or, "What support do you need from ME?"

This does not mean we are dismissing responsibility. We hold our task managers fully responsible for Active Task Management, preparation well in advance with task-level full kit, the optimal assignment of resources, and ensuring flow is the number one consideration on the task level as well as the project level.

Replacing existing local optima measurements with flow-related measurements, and replacing blame with accountability (subordination to flow) will help ensure the implementation's success. In one implementation we created a catchy slogan to communicate these insights:

"Consuming the buffer is fine. Wasting the buffer is a crime."

We can check this understanding by asking functional managers, task managers and employees, "When are you consuming the buffer?" and "When are you wasting the buffer? For example, it becomes obvious that starting without full kit and having to stop later to wait for inputs is not consuming, but wasting the buffer.

Now let's dive into the paradigm shifts in more detail - the ones that are hidden in each and every step in the S&T we have covered in this book:

5.111.1 Prioritizing and Freezing

Old paradigm: Start all projects (and all legs) as soon as possible.

New Paradigm: Start as late as possible – legs as well as whole projects.[76] Minimize bad multi-tasking and increase flow by freezing "enough" projects to lower the load on the system. Prevent starting/re-starting until defrosted, and maintain lower load on the system over time.

5.111.2 Accelerate Project Completion

Old paradigm: Spread resources across all tasks in all projects. Add resources and work overtime to meet schedules. If you have work - work. If you can open more tasks - open.

New Paradigm: Optimally assign resources to tasks in both planning and execution in priority order to increase flow and finish all projects faster. Do not assign resources to tasks/projects that are frozen (apart from preparations in 5.112.1 below) even if some resources are idle from time to time.

5.111.3 Defrost Mechanism

Old paradigm: Projects are never intentionally frozen (but task-switching is a super-freeze-generator) so they don't need defrosting.

New Paradigm: Establish simple Virtual Drum and defrost one frozen project each time one project completes the Virtual Drum phase. Keep and maintain bad multi-tasking at as low a level as possible.

---

[76] There are exceptions in unique situations. A qualified TOC consultant can advise when an exception is appropriate.

5.111.4 Releasing of new Projects

Old paradigm: Release all legs of new projects at the same time. Release new projects the minute you get them so clients see that work has begun and progress is being made.

New Paradigm: Release each new project by leg - according to the length of the legs and the rate of Virtual Drum completions. Release new projects to the system only according to the pace of the Virtual Drum and as late as possible.

5.112.1 Preparations according to priorities

Old paradigm: Start projects even though some inputs are missing.

New Paradigm: Full Kit Manager authorizes release of projects per schedule only when all preparations are (almost) complete.

5.112.2 Defining Preparations

Old paradigm: Preparations are arbitrary and undocumented.

New Paradigm: Activities which should be called preparations are officially defined, documented and followed.

5.112.3 Worried Clients

Old paradigm: Project start is almost never delayed when missing inputs. Information on delays is routinely withheld from clients.

New Paradigm: Maximum transparency and explaining of the CCPM concepts to clients enables building trust and correctly dealing with client's concerns and pressure to start early. It's not important how much has been achieved – it's important when the project will actually be delivered. Clients are informed about how preparations are valued, the CCPM system, and how the full kit policy minimizes delays in execution also how buffers protect them and that they should insist on large enough buffers within their project.

5.113.1 Building Good Project Plans/PERTS

Old paradigm: Project planning is done in many different ways and by many different people according to organization and environment. In many cases project plans are built by a single planner.

New Paradigm: There is a standard procedure (including backward planning) for project *teams* building solid project networks. It is the project team's job to plan. All critical resources are involved in planning, their wisdom is incorporated into the plans and the potential ways of accelerating execution.

5.113.2 Building Critical Chain Plans

Old paradigm: The Critical Chain is not considered.

New Paradigm: Task duration estimates are cut in half and buffers are added to projects. The team is educated that it's not important to finish all tasks on time but the project has to be delivered on time. The project buffer is of proper size and is visible to all [as well as buffer penetrations in execution]. The buffer is not hidden by top management, and belongs to the project as a whole. The team is motivated to find ways for further reduction of lead times by supporting each other and asking the simple question: "how can we go even faster?"

5.113.3 Staggering Project Portfolio

Old paradigm: Projects are started as soon as possible with minimal resources. Project delivery dates are quoted by various methods, usually a fixed lead-time from contract signing regardless of resource capacities and the existing load on the system. Clients are aware of and take actions to protect themselves against late deliveries. Mistrust is the norm between clients and suppliers.

New Paradigm: We meet our promises. Each new project is started according to the actual rate of Virtual Drum completions. Project delivery dates are quoted according to open slots in the pipeline. Peak loads on key resources are smoothed by staggering. Committed dates to clients are met with at least 95% reliability. The bottom line is we are capable of delivering many more projects, faster than before and on time.

5.114.1 Task Completion Reporting

Old paradigm: Task progress is reported by number of hours expended or percentage of work complete. In most environments the information is biased, since percentage of progress is miscalculated based on past performance. System is fed with useless and inaccurate information for the life of the project.

New Paradigm: Task status is reported every day according to days of work remaining. Expected remaining duration is the key to getting the right priorities. Regardless of the number of remaining days reported, task managers are not blamed for reporting the truth, but are encouraged to do so. Daily reporting reflects buffer penetration and enables the system to take corrective action where it really matters – and to NOT take action when there is no need to expedite. The level or urgency is by far lower, the right priorities emerge in the system, enough time to take corrective action is exposed and buffers can be protected and recovered.

5.114.2 Task Manager's role in managing execution

Old paradigm: Task managers are usually not defined as such. Those involved spend a great deal of their time in meetings and collecting and reporting task progress.

New Paradigm: Task managers review lists of active and incoming tasks, use Active Task Management, report remaining durations and delay reasons. Task managers complete task-level full kit.

5.114.3 Project Managers' role in managing execution

Old paradigm: Project managers review and manage all tasks on an (almost) daily basis. They frequently ask for status, reasons why tasks were "late", and apply pressure to task managers, resources, and others to gain an advantage for their projects.

New Paradigm: Project managers monitor and manage their project by buffer management. They are focused only on the highest priority tasks. They drive buffer recovery action plans and hold weekly buffer management meetings with the appropriate task managers, offering help as appropriate. Project managers, as well as top management, have a much better visibility and understanding of where the real problems and delays are and take corrective actions on timely manner.

5.114.4 Top Management role in managing execution

Old paradigm: Top managers, inundated with requests from all sides, are surprised by late information on project delays, react to phone calls from customers and superiors by reprioritizing frequently, and apply pressure to project managers and others for more and more detailed information. In many cases top management holds "hidden" buffers and refrains from exposing them to anyone in fear they will be used and wasted. At the same time many in the organization are aware of such buffers, and the wasting of buffers becomes the norm, as well as mistrust of management. Top management has a key role in creating more and more bad multi-tasking by flooding the system, making empty promises to clients, mandating cost savings considerations that hamper flow, and in many other ways. Flow is not the number one consideration, but rather saving costs and filling the backlog.

New Paradigm: Top managers are informed and are in full control. They hold buffer management meetings bi-weekly at a minimum, review the buffer recovery action plans and offer assistance as appropriate, and are the responsible parties for any changes to the official CCPM policies and schedule. Top management analyzes the causes for recurring buffer penetration and identify areas and policies that need to change.

5.114.5 Virtual Drum adjustments

Old paradigm: The concept of a Virtual Drum is unknown and not applied.

New Paradigm: Continuous improvement is expected and tracked, and adjustments to the Virtual Drum rate are made accordingly.

4.11.5 Mitigating Client's Disruptions

Old paradigm: Communication with clients is vague. Late inputs and change orders are often absorbed into the existing project schedule, resulting in late deliveries. Client's are often unaware of their own contribution to their late projects and blame the supplier.

New Paradigm: Clients are exposed to the concepts of CCPM and are informed of or are given access to buffer information. There is an agreed-upon mechanism in place to adjust project due dates when appropriate. Clients understand the impact of their decisions on project progress and delivery and together with the supplier agree when to insert changes and what impact those changes will have on project delivery dates. Trust is built with the supplier at levels un-known and not possible before.

4.11.6 Managing Sub-Contractors or Contracted Sub-Projects

Old paradigm: Subcontractors are often late and have all the same problems of traditional organizations - they tend to report at the last moment that they will be late. Consequently, status can be demanded frequently but answers are not always trusted. Pressure is applied on sub-contractors and the relationship tends to be authoritarian.

New Paradigm: On-going focus is provided to sub-contractors through sharing buffer information. The Company provides the right incentives for satisfactory on-time performance to sub-contractors when appropriate. The contractor greatly reduces self-caused delays to sub-contractor projects.

5.116.1 Controlling Contracted Sub-Projects

Old paradigm: Sub-contractor work is often a "black box" with little visibility and a strained relationship with the contractor, who is constantly asking for work progress reports.

New Paradigm: The requirement for sub-contractors to submit PERTs and report remaining duration daily are written into contracts. Information on project planning is shared with the sub-contractors.

5.116.2 Aligning Business Terms

Old paradigm: Since the goals of the contractor and sub-contractor are often misaligned, sub-contractors try to maximize hours worked in order to make more money.

New Paradigm: A win-win relationship exists between the contractor and key sub-contractors. The Company offers significant bonus payments (per hour) to its sub-contractors for on-time (or earlier) delivery.

A brief overview of the rest of the Projects S&T tree

We have reached the end of our initial CCPM implementation, with the S&T tree giving us a clear roadmap for ensuring the highest chance of sustainable success. But as the train arrives, we find it isn't a terminal, but only another station on the line to an ever-flourishing company. Look at the figure below. This entire book has been focused only on the activities in 3.1.1 and its Levels 4 and 5 entities (circled). This is only the **Build** phase, or the foundation for success. In order for a company to have a Decisive Competitive Edge (reliability), it must complete 3.1.2 through 3.1.5 as well. These are the **Capitalize** and **Sustain** phases. Without them, your implementation is in danger of deterioration or failure.

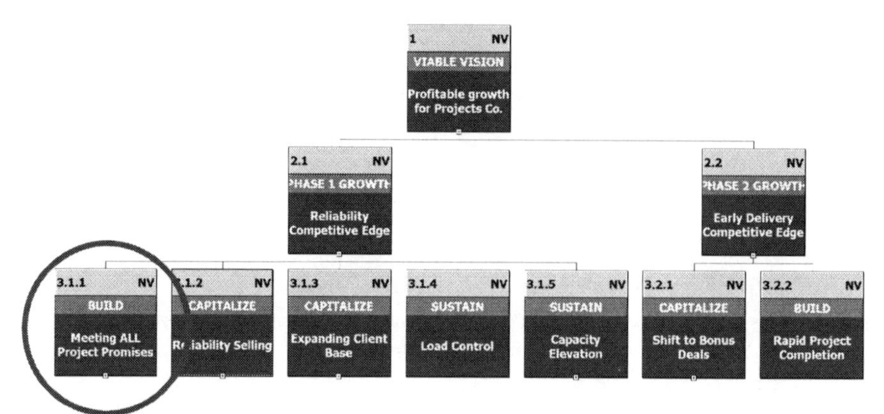

Figure 34. Everything in this book deals only with the initial Build phase of the total solution

But there's even more. Dr. Eli Goldratt often said that the true definition of an ever-flourishing company is to create a jump in improvement similar to the one we are talking about under 2.1, Reliability Competitive Edge, and then to do it *again*. Therefore once you have realized everything under 2.1, there is 2.2 (or something similar for your organization.) And every situation, no matter how much improved, can still be substantially improved even more. In the as yet uncharted future of your company, there is 2.3, 2.4 and beyond. Once you have done it twice, you "know" how to create continuous improvement – not slow, incremental improvement, but substantial, rapid jumps in improvement. Even the sky is not the limit!

All companies and all CCPM consultants should use the Projects S&T tree as the standard for implementing CCPM. Goldratt used to say that every word has been thoughtfully considered and every concept tested. It is an amazing document – a masterwork – packed full of knowledge and wisdom. **The Projects S&T tree is not a set of suggestions. It is the way – the standard – for implementing Critical Chain Project Management.** Properly implemented, Critical Chain can take your project revenue (and throughput) from this:

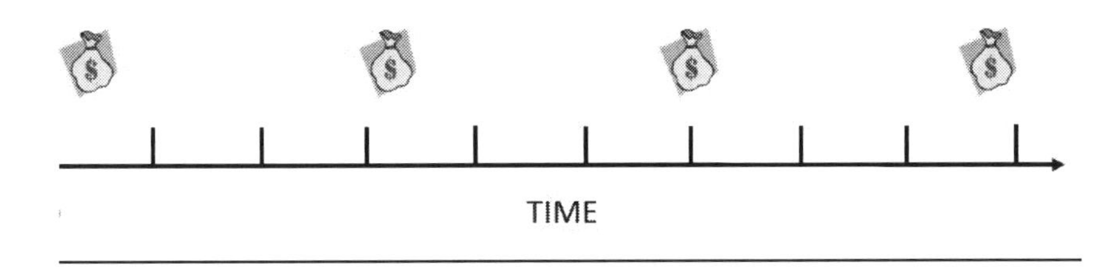

Figure 35. Project revenue stream before implementation

to this:

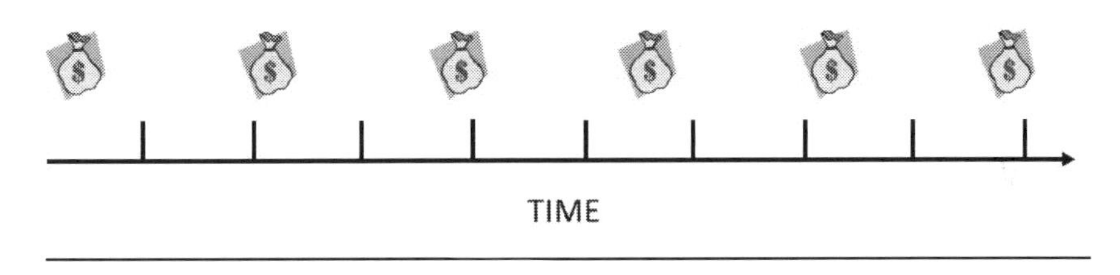

Figure 36. Project revenue stream after implementation with same resources

without any significant added expense and usually without adding any resources. However, without more knowledge, there are many dangers. Let's take a quick tour through the rest of the Projects S&T tree.

All S&T trees are constructed to secure one or more Decisive Competitive Edges. A DCE is much more than a competitive advantage. Competitive advantages have two important weaknesses – they are both *fragile* and *temporary*. For example, if your competitive edge is quality, it does not take too long for a competitor to match or exceed it. If your competitive edge is the lowest price, it can be wiped out overnight with the stroke of a pen by an aggressive competitor.

A Decisive Competitive Edge is something quite different. In the wording of 2.1 is the following statement:

"The way to have a decisive competitive edge is to satisfy a client's significant need to an extent that no significant competitor can."

DCEs are neither fragile nor temporary (over the short term). They are founded in paradigm shifts the competition has not and may not ever make. Unless they have made these shifts, your competitors are puzzled by your actions. They may even think you are foolish. Who would *offer* penalties up front for late deliveries? (See the tactics in 2.1). Projects are *always* late! Competitors will be very slow at best in responding to a Decisive Competitive Edge.

Gaining a Decisive Competitive Edge in projects begins with reliability. That is why 3.1.1, the subject of this book, is called "Meeting ALL Project Promises." But what happens if you are able to execute more projects in the same timeframe – and have not increased your sales? Instead of the result in figure 36, you could end up with something like this:

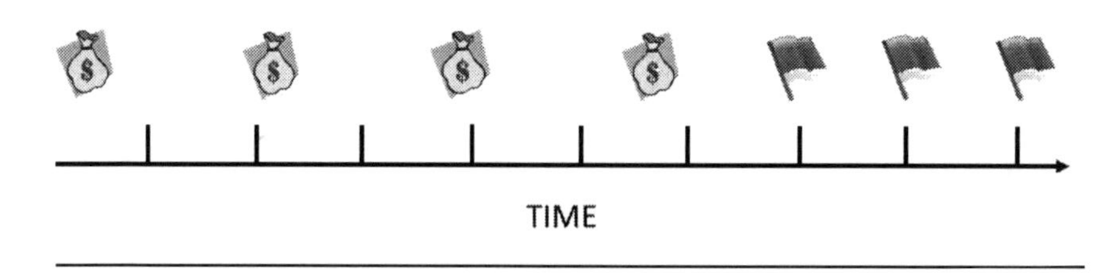

Figure 37. Project revenue after implementation without increasing sales (red flags)

Running out of work can cause serious repercussions. Cashflow could be interrupted. The implementation itself may be blamed. Peoples' jobs could be in jeopardy. As mentioned in the warnings in the "How to Use This Book" section, without taking the necessary steps to fill the sales pipeline, your increased productivity could become your downfall.

That's why entity 3.1.2 and its subordinate entities exist. In addition to your existing sales efforts, the S&T tree gives instructions on how to capitalize on your newfound reliability. It contains instructions on defining target markets, designing market offers that maximize benefits to the customer and minimize risk for both parties, and training and equipping the sales force for presenting the Reliability offer.[77]

---

[77] 4.12.4 is redundant with 3.1.3 and its subordinate entities and does not appear in all versions of the S&T tree.

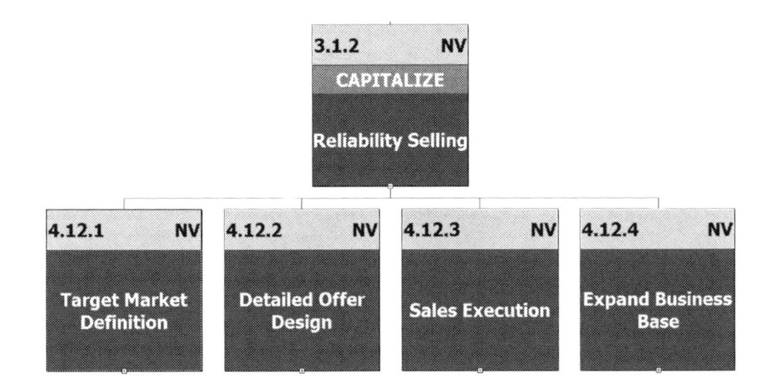

Figure 38. Reliability Selling - capitalizing on your newfound reliability

Beneath these four Level 4 entities are eleven additional Level 5 entities providing even more detailed strategy and tactics. In many ways, TOC improves upon and fundamentally changes the sales process. As with the Build phase, the Capitalization phase contains several paradigm shifts. To make these shifts successfully, the services of a qualified TOC consultant experienced in presenting marketing offers is indispensable. The value to the organization is worth many times the cost.

3.1.3 "Expanding Client Base" is another capitalization step that involves generating new leads based on the Reliability Offer, and monitoring and managing the sales pipeline.

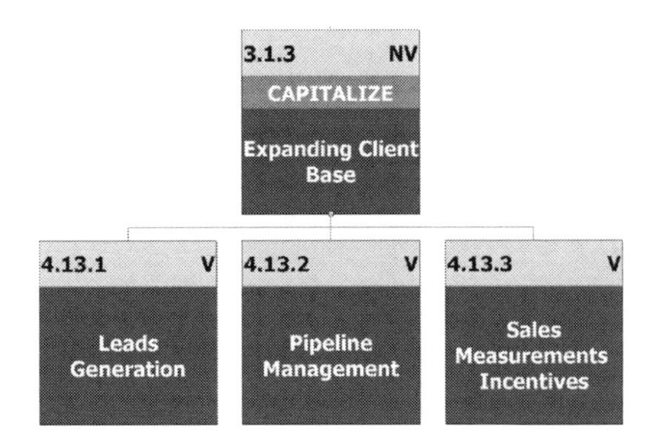

Figure 39. Expanding Client Base - further capitalizing on the reliability offer

The Level 4 entities beneath 3.1.3 outline more specific strategy and tactics for developing a mechanism for generating leads while using less of the salespeople's capacity, smoothing the load and maximizing the throughput of the sales pipeline, and providing the right incentives for salespeople – in alignment with the holistic goals of the company.

Figure 40. Load Control - Sustaining the implementation during increases in sales

3.1.4 Load Control is a Sustain step that involves how to maintain on time delivery and avoid overload during rapid increases in sales volume. This is a crucial step for any implementation that contributes to the accelerated growth of an organization. Without an infrastructure prepared for such growth, the organization is in danger of returning to chaos and poor delivery performance, or even of collapse and failure. Goldratt illustrated the situation in the following graph, which we call the "Green and Red Curves":

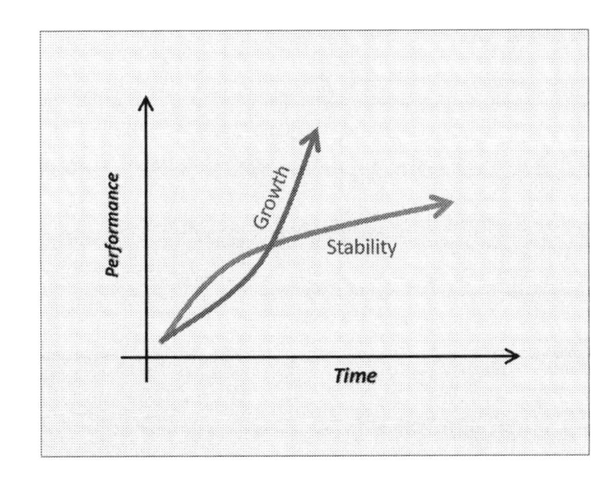

Figure 41. The Red and Green Curves of Growth and Stability

The key learning from this graph is that in order to accommodate rapid growth (the red curve), the organization must also be increasing its stability (the green curve) at a sufficient rate. Actions must be taken to ensure that growth does not get ahead of stability. Therefore the strategy and tactics must include the necessary steps for accomplishing this.

3.1.5 Capacity Elevation and its Level 4 and 5 detail steps are another set of Sustain steps that help keep the company on the green curve while allowing for growth on the red curve. In TOC, the word "Elevate" is synonymous with "get more" – in this case, getting more *capacity.* The S&T tree takes two approaches for getting more capacity. The first involves internal improvement – getting more capacity through better managing the existing system – or finding "hidden" capacity (which is hidden by our inefficiencies) inside our system. Just as the initial implementation provided capacity, we can find more over time.

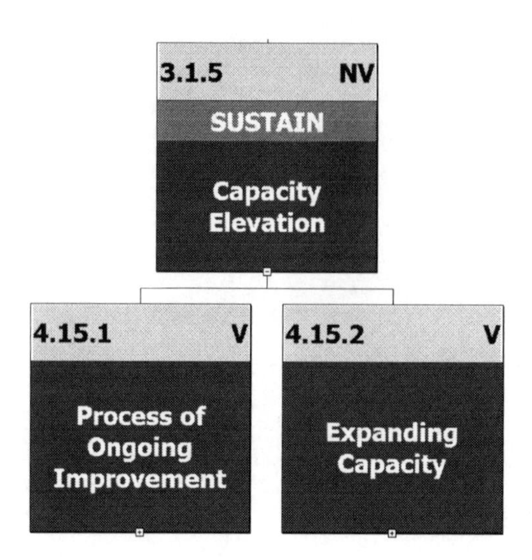

Figure 42. Strategy and tactics for getting more capacity

4.15.1 Process of Ongoing Improvement (or POOGI) and its three subordinate entities provide a method for collection and analysis of relevant delay reasons in projects, and offers a plan for deploying improvement teams.

The second approach, 4.15.2 Expanding Capacity, is about acquiring additional capacity in a more traditional, but also much more focused sense. Its two subordinate entities concern identification of resources which may need elevation, and the acquisition of resources relative to their expected sequence and timing of need.

As was mentioned in Chapter 2, S&T trees can be customized to fit an organization as necessary under the guidance of a qualified TOC expert. Usually the Build phase under 3.1.1 is followed *without modification*, since reliability is important to virtually all organizations, even if that reliability is limited to internal commitments – meeting our promises to ourselves. Outside of 3.1.1 customization is more common. But for many if not most Projects organizations, reliability to external customers is a high priority. In such cases it also often makes sense to define a second Decisive Competitive Edge to be implemented upon the successful completion of the first: 2.2 Early Delivery Competitive Edge.

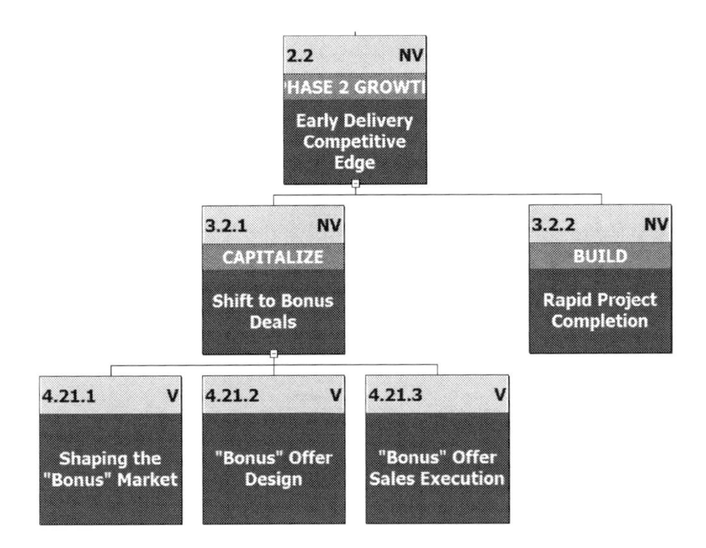

Figure 43. Early Delivery - A second Decisive Competitive Edge (DCE)

The objective of the Early Delivery DCE is to capitalize further on the new capability of the company. Especially in environments where time is of high financial importance, such as in oil and gas plant construction and turnarounds – where gained or lost production can be worth millions of dollars per day, the shortening of lead time can lead to premium pricing or bonuses. This is why I cautioned you in Chapter 24 not to run too fast to promise your clients shorter lead times. Sometimes reliability itself is enough, and just being reliable to existing market standard lead times will bring in significant new business. The shorter lead times are often a plus for you, and can be turned into new market offers that will further increase your throughput.

Even if 2.2 is not applicable to your organization, there is a likelihood that a qualified TOC expert can find for you other DCEs. The power and potential of the S&T tree is amazing. Its limits, if any, have not been tested. Now maybe you can see why Dr. Eli Goldratt called the S&T trees the most important thing he ever developed.

Maybe you can now also see why the Projects S&T tree is so effective as a vehicle for implementing Critical Chain Project Management – why it gives you the best chance for success and sustainability. Maybe you'll consider it for use in your implementation. Even if you have already implemented, and aren't seeing 100% of the benefits hoped for or possible, you'll consider an audit of your work based on the Projects S&T tree, and attempt to close any gaps you may discover.

Your future can be ever-flourishing. Join the many companies that have started on this journey. Even the sky is not the limit. I wish you the best of luck.

Oh, and one more thing...

Flow is the number one consideration.

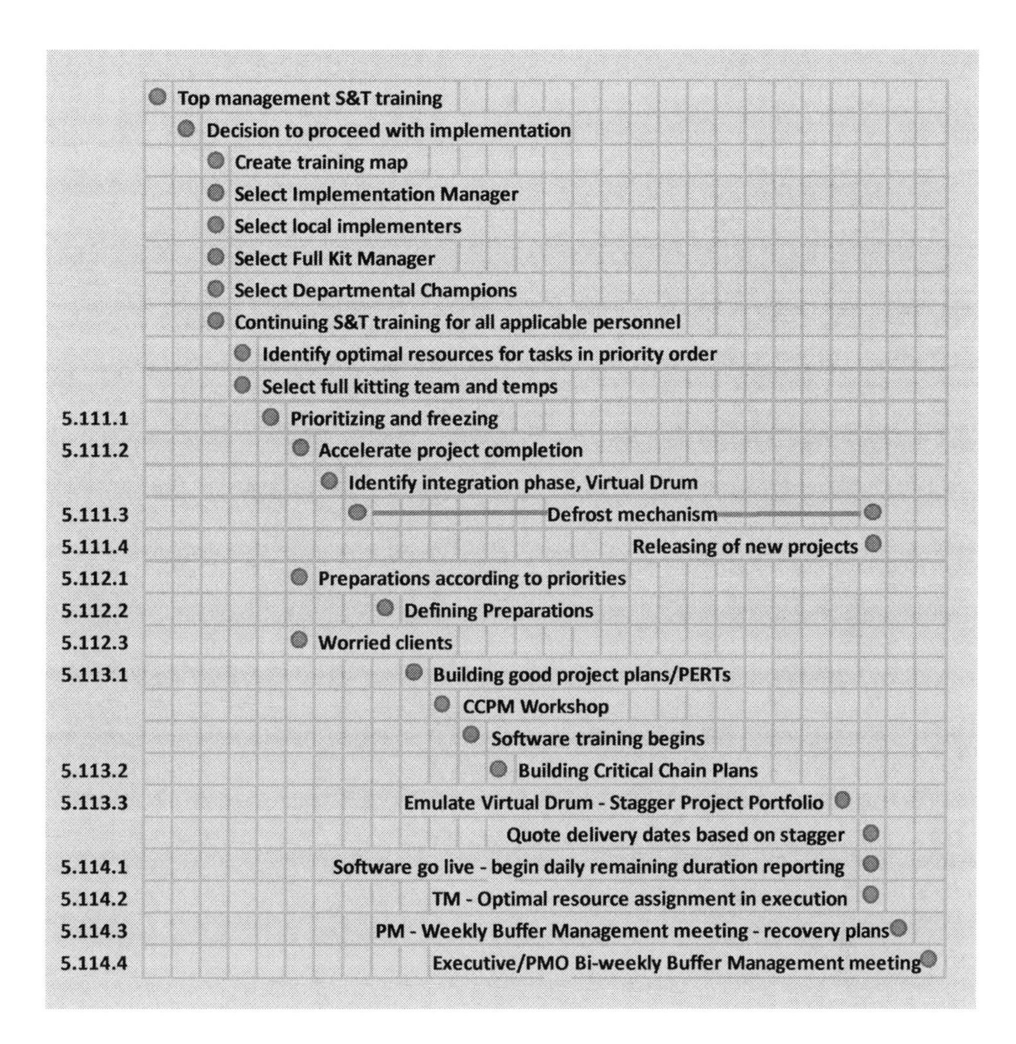

Figure 44. Implementation highlights – Approximate timing

About the Author

DAVID UPDEGROVE is a Theory of Constraints implementation expert, certified in all TOC disciplines including Production, Project Management, Supply Chain, Finance & Measures and Business Strategy. He has 26 years TOC-based Performance Improvement experience, and has led or participated in some of the most successful and enduring TOC implementations in the world.

He currently serves as Senior Advisor to the TOC Think Tank of Nagoya, Japan, and is on the Board of Directors of Progressive Flow, Ltd., a TOC consultancy based in Israel.

Printed in Poland
by Amazon Fulfillment
Poland Sp. z o.o., Wrocław